GAS SERVICE TECHNOLOGY 3

THIS BOOK is the third and last of a series of manuals devoted to the theory and practice of gas service. The volumes are the results of some years of investigation and planning, first by a Working Group of the then Gas Council and subsequently by the Gas Manuals Steering Committee of the British Gas Corporation.

The Working Group established the need for the three volumes, to deal with the basic science and practice of gas service, domestic installation and servicing practice, and commercial and industrial servicing practice. The Steering Committee then undertook the task of producing the manuals. The detailed scheme of contents of each of the volumes was originally drafted by the Committee's Coordinating Editor, Ken Pomfrett, who was delegated the task of commissioning the authors and assembling their contributions. The technical editing was carried out by George Jasper, Senior Training Adviser (Customer Service) who recast the individual contributions into consistent style as a coherent and comprehensive body of information.

Gas Manuals Steering Committee

PETER ANDREWS* *Chairman*

GEORGE JASPER *Editor*

KEN POMFRETT *Coordinating Editor*

Original Members

G. BATTISON
Editor-in-Chief,
Gas World Group,
Benn Publications Ltd

E. W. BERRY,
Senior Scientist,
Watson House, BGC

G. S. E. HOPPER *Secretary*
Assistant Education and Training
Manager, North Thames Gas

K. POMFRETT
Tan-y-Craig Farm,
Llanfair Caereinion

L. J. SMILLIE
Assistant Training Officer (Service),
North West Gas

F. WEBB
Assistant Chief Service Engineer
Service Department, BGC

B. J. WHITEHEAD
Deputy Head of Construction and
Services Department, Cauldon
College of Further Education,
Stoke-on-Trent

N. J. URWIN†
Head of Manpower Resources and
Development, Personnel Division,
BGC

Later Members

P. J. ANDREWS* *Secretary*
Senior Training Adviser (Customer
Service), Personnel Division, BGC

R. D. BRUCE
Services Manager,
Watson House, BGC

M. S. GALE,
Editor (Technical Books),
Ernest Benn Ltd

E. GLENNON
Assistant Training Officer (Service),
North West Gas

D. LEWIS
Chairman,
Gas Teachers Association

G. W. JASPER ‡
Senior Training Adviser
Personnel Division, BGC

*Chairman from 1976 to 1984
†Chairman from 1972 to 1974
‡Chairman from 1974 to 1976

Gas Service Technology 3

Commercial and Industrial Gas Installation and Servicing Practice

Edited by George Jasper

Coordinating Editor Ken Pomfrett

Published in association with
the British Gas Corporation by
BENN TECHNICAL BOOKS

First published 1980 by Ernest Benn Limited

This impression published by
Benn Technical Books
Tolley House, 17 Scarbrook Road, Croydon CR0 1SQ

First edition 1980
Second impression 1981
Third impression 1982
Fourth impression 1984

Typeset by Reproduction Drawings Ltd

Printed and bound in Great Britain by
Anchor Brendon Ltd, Tiptree, Essex

British Library Cataloguing In Publication Data
Gas service technology.
  3: Commercial and industrial installation and servicing practice
  1. Gas appliances – Maintenance and repair
  I. Jasper, George   II. British Gas Corporation
  III. Pomfrett, Ken
  665'.75   TH6860

ISBN 0-510-47440-3

Contents

Preface

Since the introduction of natural gas in 1967, its use in the industrial and commercial markets has increased about eightfold. Many processes have changed over to gas firing, attracted by its cleanliness, versatility, controllability and freedom from delivery and storage problems. This has resulted in an increase in the servicing workload, and more men are becoming involved with industrial and large commercial equipment.

This volume has been re-cast with this in mind. It attempts to deal in a general manner with the use of the equipment and with industrial and commercial installation and servicing practice. It also contains some more advanced related theory to help the student understand clearly the practical applications described.

A considerable amount of work has taken place in the development of equipment and control devices, resulting in reports and technical notes being produced by British Gas and allied organisations. These notes are an invaluable source of information and they have been referred to where they are relevant. However, because development is continuous and revisions are frequent it is possible that some of the information in this volume may be out-of-date by the time the book comes to publication. Reference should always be made to current codes of practice, reports and notes for guidance.

Acknowledgements

In addition to those mentioned in the two previous volumes, the following have helped to revise and check Volume 3.

Mr. Martin C. Pascal, Senior Service Officer (Engineering), British Gas HQ, together with a number of his colleagues, has checked most of the scripts for technical accuracy.

Mr Jim Cornforth, Tutor, School of Fuel Management; he not only revised his own original contribution but also supplied up-to-date information for other chapters.

In addition to the technical assistance received, I am also indebted to my secretary, Mrs. Jo. Kippen who typed and circulated the scripts and revisions for most of this volume and finally to my wife, Grace, without whose support and encouragement the Manual would not have been brought to a satisfactory conclusion.

G. W. Jasper
M.B.E., C. Eng., F.I. Gas E.

CHAPTER 1

Large Installations

Chapter 1 is based on an original draft by Mr. E. Glennon

§1 Introduction

Large diameter pipework in non-domestic premises should be installed to comply with the requirements of the Gas Safety Regulations and the recommendations of the Standard and Installation Code of Practice for gas pipework for Industrial and Commercial Applications, Report IM/106.

This code is intended to supplement and to complement the existing B.S.C.P. 331 Part 3. It deals with the materials and methods for installing pipework downstream from the meter control to the burners on the equipment. It applies to pipes of 15 mm ($\frac{1}{2}$ in) or larger diameter for pressures up to 5 bar (72 lbf/in^2).

Work on large installations poses a number of problems. Some of these are also associated with the smaller pipework in domestic and commercial premises and have been dealt with in Volume 2.

Pipe sizing is generally carried out as described in Volume 1 although gas flow calculators are available and offer a convenient method of calculating gas flows, velocities and pressure drops.

Velocities can be critical, particularly in unfiltered supplies where dust or debris at high velocities can cause erosion or damage to valves or controls. Generally on supplies filtered to $250\,\mu$m a maximum velocity of 45 m/s (148 ft/s) is permissible.

Pipes are usually mild steel, copper, ductile iron or polyethylene (P.E.). In high rise buildings steel pipe is recommended for all vertical risers above 3 m (10 ft) and all pipes above 50 mm (2 in) diameter.

§2 Pipework Layouts

The route selected for the pipework should be as short as possible without being too obtrusive. The design should allow for possible future extensions.

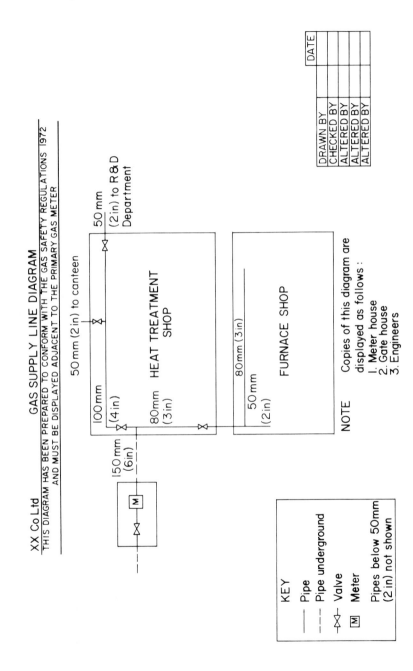

Fig. 1. Gas supply line diagram

The installation should include valves as necessary to provide:

- section isolation
- purging
- use in emergency

For commercial or industrial premises having two or more floors to which gas is supplied by a service pipe larger than 50 mm (2 in) diameter, valves must be fitted to enable each floor to be isolated. Where a single floor is divided into self-contained areas, the pipework to each area must also be valved.

In addition, a line diagram should be provided to enable isolating valves to be quickly located in case of emergency. One copy of the diagram must be fitted as near as possible to each primary meter. Other copies may be placed at the gate-house and the services engineers' office as appropriate. The diagram must be up dated whenever alterations are made to the installation.

An example of a supply diagram is shown in Fig. 1. It shows sufficient detail to identify the isolating valves but need not include every final connection. A key should always be provided. Diagrams are usually A4 size and protected by glass or plastic. The provision of the diagram is the responsibility of the installer and the occupier, in the case of a factory.

§3   Pipework Buried Underground

Supplies running from one building to another may be buried or carried in ducts. A typical case is where the meter house is isolated from the main building. Where installation pipework is buried it should generally conform to the recommendations for service pipes (Volume 2, Chapter 4) and reference should be made to relevant publications and standards. These include the I.G.E. publication No. TD4. "Laying of Steel and Ductile Iron Gas Service Pipes".

All pipes must be adequately protected against corrosion. Up to 50 mm (2 in), pipes are obtainable with factory applied wrapping or sheathing. Joints or exposed sections of pipe must be covered, after testing for soundness, usually with self adhesive PVC tape or a bandage impregnated with a petroleum grease. Larger, uncoated pipes must be wrapped after being laid. On some sites cathodic protection may be necessary or the excavation may be filled with a passive material, chalk or sand.

Exposed pipework may be painted rather than wrapped unless situated in a very corrosive atmosphere.

The route for underground pipework must be selected with the following points in mind:

- it must avoid close proximity to unstable structures
- it must be kept clear of walls which retain materials above the level of the ground in which the pipes will be laid
- pipes must not pass under load-bearing walls, foundations or footings
- pipework must not be laid through large unventilated voids
- areas where there may have been a recent infill must be avoided unless steel pipe for pressures up to 5 bar ($72\,\text{lbf/in}^2$) or ductile iron with locking joints for pressures up to 2 bar ($30\,\text{lbf/in}^2$) is used.
- pipes operating at pressures between 2 and 5 bar (30 to 72 lbf/in^2) should be laid at least 3 m (10 ft) from any building until turning to enter that building.

The amount of cover above the pipe must be at least that given in Table 1. In special circumstances this may be reduced if the back fill is suitably reinforced.

§4   TABLE 1   Depth of Cover on Buried Installation Pipes

	Minimun Depth of Cover	
Pipe Size	Roadways and Grass Verge	Paved Foot Walk
50 mm (2 in and below)	375 mm (15 in)	375 mm (15 in)
Above 50 mm (2 in)	750 mm (30 in)	600 mm (24 in)

Where valves are fitted underground they should be provided with access to spindles and lubrication points. Valve pits should have covers of adequate strength or surface boxes to BS 1426. The position of the valve should be indicated by a marker plate, Fig. 2. This plate is also used to show the position of syphons or purge points.

Although syphons or dip pipes are not necessary on natural gas supplies, they may be fitted to pipework which will be hydraulically tested to provide a means of removing the water.

On very congested sites it is advisable to fit markers in the road surface to show the route of the gas installation, particularly where there is a change of direction or intersection. A typical marker is shown in Fig. 3.

V denotes valve
S denotes syphon
PP denotes purge point
I3 denotes distance of marker plate from V, S or PP
(At present feet is the standard unit)

Minimum dimensions 150 mm x 150 mm

Fig. 2.   Marker plate; to show valves, syphons or purge points

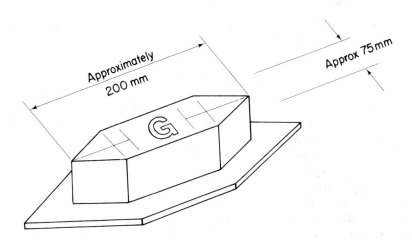

Fig. 3.   Pipe route marker

§5 Pipework Above Ground

Pipes entering buildings should pass through the wall or floor in sleeves sealed at each end by a non-hardening, non-combustible material to prevent the passage of water, vermin or gas.

The dimensions of suitable sleeves are given in Table 2.

§6 **TABLE 2 Pipe Sleeves**

Pipe Bore	15 mm ($\frac{1}{2}$ in)	20 mm ($\frac{3}{4}$ in)	25 mm (1 in)	32 mm ($1\frac{1}{4}$ in)	40 mm ($1\frac{1}{2}$ in)
Sleeve Bore	25 mm (1 in)	32 mm ($1\frac{1}{4}$ in)	40 mm ($1\frac{1}{2}$ in)	50 mm (2 in)	65 or 80 mm ($2\frac{1}{2}$ or 3 in)

Pipe Bore	50 mm (2 in)	80 mm (3 in)	above 80 mm (3 in)	
Sleeve Bore	80 mm (3 in)	100 mm (4 in)	annular space not less than 13 mm ($\frac{1}{2}$ in)	

Pipework should not be laid through electrical intake chambers, transformer rooms or lift shafts. It should be spaced at least 25 mm (1 in) away from any electrical or other service. On large diameter installations spaces up to 250 mm (10 in) may be required.

The pipes should be adequately protected against corrosion both internally and externally. Precautions should be taken to prevent the entry of dirt, debris and welding scale into the pipework during installation.

Gas pipes should be electrically cross bonded to other services as described in Volume 1, Chapter 8. When any pipes are disconnected, a temporary continuity bond should be attached before the supply is broken (Volume 2, Chapter 3).

If pipework is exposed in a high position, for example on the roof of a tall building, it should be protected by fitting lightning conductors to BS CP 326.

Gas pipework should be easily identifiable in accordance with BS 1710. Where there are no other piped gas supplies it is sufficient to paint the pipes with Yellow Ochre 08C 35 to BS 4800. Where there are other piped gases, for example on a chemical works, it is desirable to provide more precise indentification by a secondary colour band over the base of yellow Ochre. For natural gas the secondary band colour is Yellow 08 E51. Alternatively, the name of the gas or its chemical symbol may be used.

§7 Pipework in Ducts

Ducts which carry gas installation pipes should have through, natural ventilation to prevent an accumulation of gas. There should be at least two ventilation grilles to outside air, one at each end of the duct. These should give a ventilation rate of at least $0.7 \, \text{m}^3/\text{h}$ ($25 \, \text{ft}^3/\text{h}$). In exceptional cases higher ventilation rates and additional grilles may be necessary.

Other services at pressures below atmospheric or ventilation ducts should not be installed in the same duct as a gas supply, to avoid any infiltration of gas.

All pipes which pass through the walls of a duct must be fire-stopped. Pipes above 150 mm (6 in) diameter should not pass through the walls of a duct.

Access into ducts must be provided for servicing purposes but the number of openings should be kept to a minimum. Access panels must not open on to any:

- fire exit route
- emergency stairs
- lift landing

Panels must have $\frac{1}{2}$ hour fire resistance and conform to BS 476 Part 8. Although generally gas pipework should not pass through unventilated ducts or voids, pipes may be installed when:

- the pipe is run through the space in a continuous sleeve ventilated at one or both ends
- the space is filled with an inert, dry, fire resistant material

Ducts should be checked, at least annually and also before and after carrying out any work on the pipework, for any trace of gas leakage. A portable leak detector should be used.

§8 Jointing

Steel

Steel pipework for normal working pressures can be screwed or welded for use in ducts up to 100 mm (4 in) diameter and above ground up to 150 mm (6 in). Larger diameters and pipes at higher pressures must only be welded.

Where pipe is screwed to BS 21, malleable iron fittings are used as described in Volume 2 Chapter 1. Alternatively, flanges may be fitted either by screwing or welding on to the pipe. Where pipework is welded the number of flanged joints should be kept to a minimum and they should all be welded on to the pipe.

Flanges should conform to BS 4504, Part 1. Flanges to BS 10 are

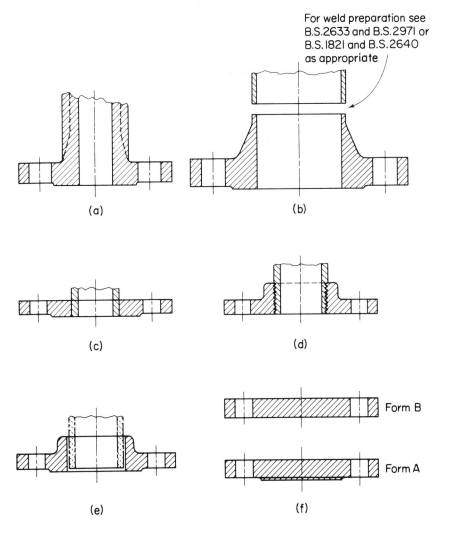

Fig. 4. Flanges (BS 4504): (a) integral flange (b) welding neck (c) plate flange (d) C.I. screwed boss (e) slip-on boss (f) blank flanges

obsolescent although many are still in use on the district. There are a number of different types shown in Fig. 4. These include:

a) integral flanges—part of a valve or fitting
b) welding neck—for welding to pipe end
c) plate—for welding to pipe wall
d) screwed boss—for screwing to threaded pipe
e) slip on boss—for welding to pipe wall
f) loose—for welded on lapped pipe ends (Form A)
   blank—for sealing off a pipe or fitting (Form B)

Gaskets for flanges may be either "full-face" or "inside bolt circle" (IBC).

Full-face gaskets extend to the outer edge of the flange and are used on flanges which have a perfectly flat face, Fig. 4 (d).

IBC gaskets have an outside diameter equal to the bolt pitch circle less one bolt hole diameter. They are used on flanges with a centre raised section, Fig. 4 (a).

Typical flanged fittings for use up to 14 bar (200 lbf/in^2) to BS 10 are shown in Fig. 5.

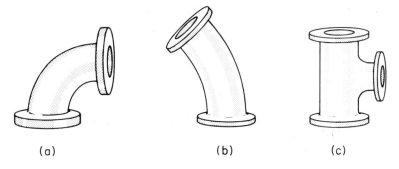

(a)　　　　　　　　(b)　　　　　　　　(c)

Fig. 5. Typical flanged fittings

Semi-rigid pipe couplings and adaptors may be used as an alternative to screwed or welded joints provided that they are accessible. Compression couplings should not be used in ducts in high rise buildings.

Copper

On copper tube, capillary and compression fittings are used (Volume 2, Chapter 1) but only on tubes up to 42 mm (1½) outside diameter.

Compression joints may only be used where they are accessible, they may not be enclosed in a sleeve or a sealed duct.

Capillary and compression joints may not be used in vertical ducts in high rise buildings.

Polyethylene (P.E.)

This pipe is extensively used by the Distribution Department and is described in Volume 2, Chapters 1 and 4. It is only suitable for use below ground, because it deteriorates on exposure to daylight.

Jointing is commonly by compression couplings with reinforcing inserts or by fusion welding. Fusion welding requires the use of special tools by trained operatives.

§9　Associated Components

Valves

Manual valves are fitted to isolate different floors and, in some cases, each self-contained area. They should be fitted into pipework before it enters the building and at branch off take points.

Valves should also be fitted for plant isolation upstream of all other plant controls. They should be provided with a means of disconnection, for example a union or flange, immediately on their outlet.

All valves should be located with adequate access for lubricating and servicing. Automatic valves and controls should be fitted in accordance with manufacturer's instructions and mounted securely and level in non-vulnerable positions. The fitting of non-return valves is dealt with in Chapter 4.

Filters

Where there is a likelihood of dust causing erosion or adversely affecting the operation of the plant a filter should be fitted. This would normally remove particles larger than $250\,\mu$m although in some cases finer filters may be necessary.

Filters should be fitted immediately upstream of the plant to be protected. Where the plant is not more than 20 m downstream of a meter governor filter additional filters may not be necessary.

Purge points

Plugged or capped purge points should be fitted at isolating valves and in other appropriate positions to allow the installation to be safely tested and purged in accordance with the recommended procedures. Each purge point should normally incorporate a valve to control its operation.

§10   Pipe Supports

Pipework must be adequately supported to BS 3974. Owing to their considerable weight, large diameter pipes require much stronger supports than smaller installations. In factories pipework may be subjected to extremes of temperature and vibration, both of which may cause movement. The coefficient of linear expansion for mild steel is 0.011 mm/m degC, so the supports must allow for thermal movement without damaging any insulation or corrosion protection applied to the pipework.

The spacing of the supports is given in Table 3. However, where hydraulic soundness testing is to be carried out allowance must be made for the additional weight of the water. Hydraulic testing is used on pipework operating at pressures above 2 bar (30 lbf/in^2).

§11   **TABLE 3**   **Maximum Distances Between Pipe Supports**

| Pipe Size | Maximum Distance Between Supports (m) | | | | | |
| | Horizontal Pipework | | | Vertical Pipework | | |
mm (in)	Cast Iron Steel (not Welded)	Steel (Welded and Flanged)	Copper	Cast iron Steel (not Welded)	Steel (Welded and Flanged)	Copper
15($\frac{1}{2}$)	2.0	2.5	1.2	2.5	3.0	2.0
20($\frac{3}{4}$)	2.5	2.5	1.8	3.0	3.0	2.5
25(1)	2.5	3.0	1.8	3.0	3.0	2.5
32($1\frac{1}{4}$)	2.7	3.0	2.5	3.0	3.0	3.0
40($1\frac{1}{2}$)	3.0	3.5	2.5	3.5	3.5	3.0
50(2)	3.0	4.0	2.7	3.5	3.5	3.0
80(3)	3.0	5.5	3.0	3.5	4.5	3.5
100(4)	3.0	6.0	3.0	3.5	4.5	3.5
150(6)	4.0	7.0	NR	NR	5.5	NR
200(8)	NR	8.5	NR	NR	5.5	NR
250(10)	NR	9.0	NR	NR	6.0	NR

NR = Not recommended.

Generally the structure of industrial building lends itself to the carrying of pipe supports. Brackets and hangers can be attached to rolled steel trusses, girders and columns by lugs, bolts or welding. On concrete sections, supports may be secured by masonry bolts or fitted to plates and held by bolts through the structure.

Horizontal pipework

Rollers and chairs mounted on brackets are a simple method of supporting uninsulated pipe while allowing lateral movement. Examples are shown in Fig. 6. Rollers are usually cast iron although small sizes are available in bronze.

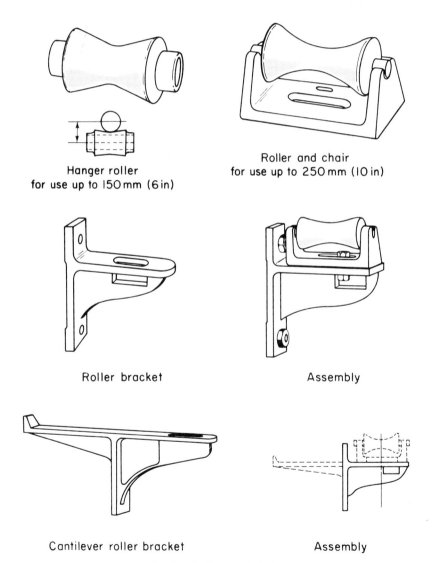

Hanger roller
for use up to 150 mm (6 in)

Roller and chair
for use up to 250 mm (10 in)

Roller bracket

Assembly

Cantilever roller bracket

Assembly

Fig. 6.   Rollers and chairs

For insulated pipes, sliders or cradles with rollers and base plates may be used, Fig. 7. The pipe is first clamped to the cradle which rests on a flanged roller mounted on a flanged plate. When positioned the pipe is insulated.

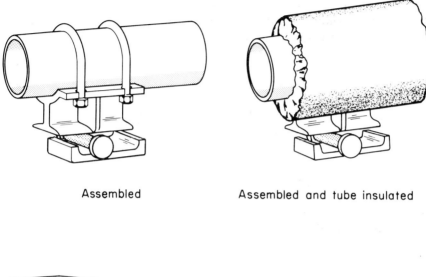

Assembled   Assembled and tube insulated

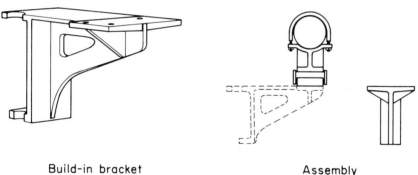

Build-in bracket   Assembly

Fig. 7.   Cradle and roller

Where banks of pipes are run horizontally together they may be supported by low level trestles, Fig. 8, or suspended from a roof truss or girder as in Fig. 9. In both these examples shown, rollers have been dispensed with. The pipes are held in position by flat guide plates and slide directly on the supporting steel section.

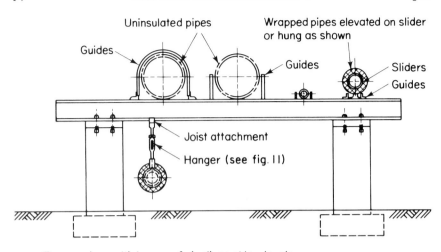

For carrying multiple runs of pipelines at low levels.

Fig. 8.   Low-level trestles

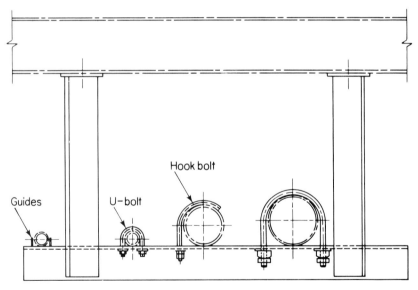

This simple form of support, attached below another structure, enables banks of pipe to be carried in the horizontal plane.

Fig. 9.   Suspended support

Typical pipe support brackets which may be used for single or multiple runs of pipe are shown in Fig. 10. The diagram illustrates the methods of securing the brackets to the structure.

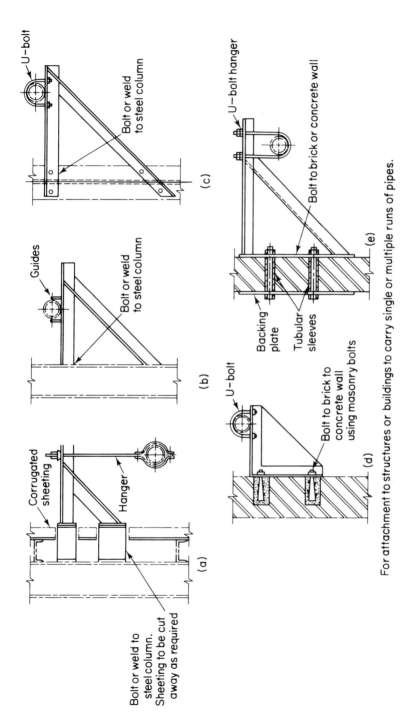

Fig. 10. Pipe support brackets: (a), (b), (c) bolted or welded to steel column (d) masonry bolts (e) bolted through concrete column

For attachment to structures or buildings to carry single or multiple runs of pipes.

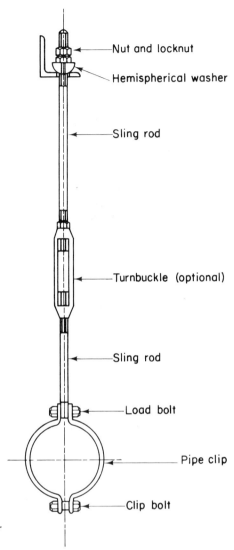

Nut and locknut

Hemispherical washer

Sling rod

Turnbuckle (optional)

Sling rod

Load bolt

Pipe clip

Clip bolt

Fig. 11. Hanger for uninsulated pipe

A simple hanger for uninsulated pipes is shown in Fig. 11. Because it is suspended by a domed washer, the sling rod has a limited amount of pivoting movement to allow for small movement of the pipework. A more complex, spring loaded hanger suitable for insulated pipe is shown in Fig. 12. This can accommodate positional changes in a pipe. The spring housing carries the full load on the hanger and the hanger should only be used where load changes are relatively small.

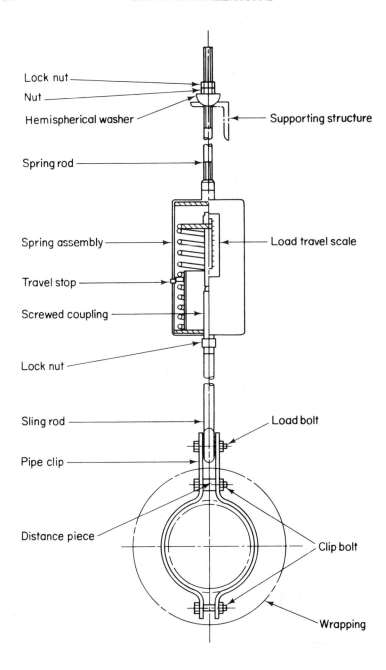

Fig. 12. Spring-loaded hanger for insulated pipe

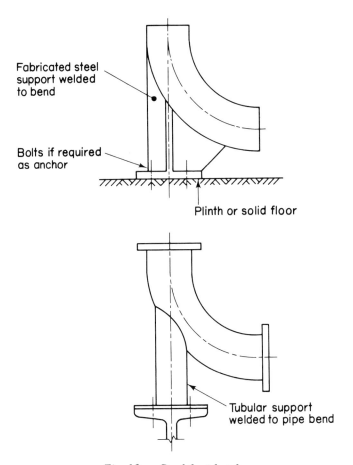

Fabricated steel
support welded
to bend

Bolts if required
as anchor

Plinth or solid floor

Tubular support
welded to pipe bend

Fig. 13.   Duckfoot bends

Vertical pipework

Vertical pipes may be supported from:

- below, by duckfoot bends
- on the run, usually by spring hangers
- the top of the riser, by hangers allowing some free movement.

Duckfoot bends are secured to the pipe line by welding or flanges, Fig. 13. They are fitted at the bottom of a riser and are often mounted on a concrete plinth which may be separate from the main structure.

Spring hangers, Fig. 14, support the pipework by spring loaded sling rods attached to brackets which have been welded to the pipe wall. The pipework may be insulated or wrapped after the brackets have been fixed.

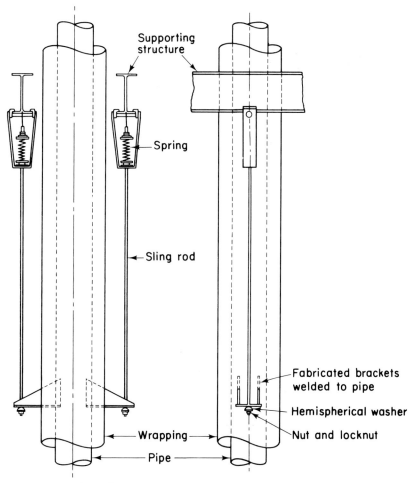

Fig. 14.   Spring hangers for vertical pipe

Figure 15 shows two alternative methods of supporting vertical pipework. They may both be used with insulated or uninsulated pipe and offer some limited free movement. The hanger at (b) is generally more suitable for heavier applications.

§12   Fabricating Pipework

On large installations the fabrication of pipework is extensively practised in workshops away from the actual installation. Fabrication may be carried out either by the manufacturers of pipes and fittings or by the installer.

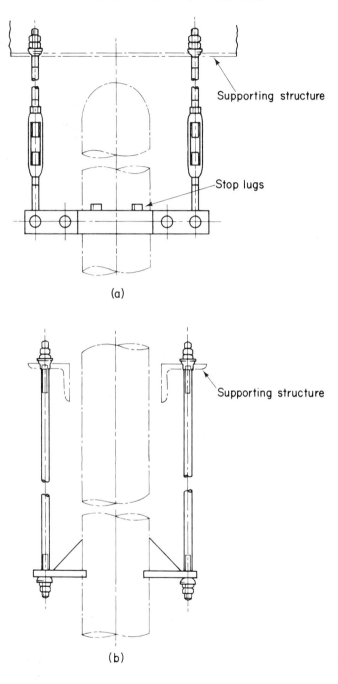

(a)

(b)

Fig. 15.   Hangers for vertical pipes: (a) hanger with some limited free movement (b) support for heavier riser

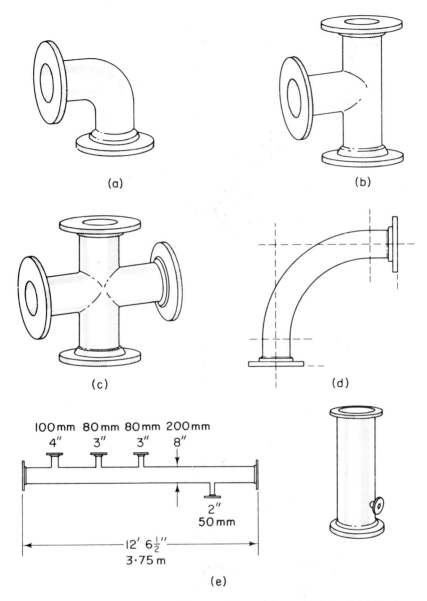

Fig. 16.   Fabricated fittings: (a) elbow (b) tee (c) cross (d) bend (e) header

Manufacturers offer a range of items, examples of which are shown in Fig. 16. The elbows, tees, crosses and bends are made to standard sizes and reducing fittings are available. They are fabricated usually from mild steel pipe with flanges welded and drilled.

Headers or manifolds can be obtained to any pattern or size required. They are useful where a number of branches are needed close together.

Installers, including those regions of British Gas which serve large industrial areas, often use mobile workshops. These enable pipework with awkward sets to be made up in sections and screwed or welded and flanged ready for assembly on site. The mobile workshops carry similar equipment to that used in base workshops. Portable electric welding plant can be carried from the vehicle to the assembly site and may be included.

§13 Flexible Connections

Pipework in industrial situations may be subjected to:

- movement
- expansion
- vibration
- strain

Because of this, some means must be found to counteract these forces and flexible or semi-rigid connections may be used.

Most large buildings are prone to some structural movement and high rise buildings, for example, are designed to sway slightly with wind pressure. So allowance must be made for the installation pipework to move relative to the incoming supply. Corrugated bellows-type steel connections are often used on lateral offtakes from vertical risers.

Thermal expansion is most likely to have its greatest effect on long lengths of exposed pipework or connections to furnace burners. Underground pipework is not normally affected. Semi-rigid couplings can accommodate lateral movement up to 10 mm. They should only be used as expansion joints if all joints are couplings. Flexible bellow-type connections are satisfactory if fitted with restraining ties.

Vibration is normally due to moving machinery, for example, gas compressors or mixers, gas engines or automatic process machinery. To prevent vibration being transmitted throughout the installation pipework, armoured flexible metallic tubing is commonly used. Couplings are unsuitable for continuous vibration.

Strain may be produced in a number of ways. Where it is likely to be caused by misalignment of final connections, couplings may be used. Two couplings joined by a short length of pipe can accommodate a lateral displacement.

Where strain is due to torsion or rotation it is best prevented by the use of swivel joints.

Flexible connections in the form of sheathed flexible tubing are generally used where burners or appliances must be moved to gain access for servicing or cleaning. They can also connect supplies to floating restaurants or bars moored in tidal waters. Where equipment is used in several different locations, flexible connections can correct any misalignment at each point of use.

When using flexible connections the following points should be noted:

- current codes of practice should be studied to determine the circumstances under which each form of connection may be used
- the length of the connection should be as short as possible
- it must not pass through any rigid structures, for example, walls or floors
- only couplings may be used in buried pipework
- connections should otherwise be in accessible locations
- it must be protected against heat and mechanical damage
- it must be checked for leakage at regular intervals
- where electrical continuity is required it may be necessary to fit separate bonding across semi-rigid couplings or flexible tubing
- where couplings are fitted above ground they should be provided with some form of restraint to prevent the pipes separating, tie rods or chains may be used
- individual pipes must be securely anchored to ensure that the flexible connections are not strained
- there should be an isolating valve upstream of the flexible connection

§14 Tools

Pipe gripping tools

These are generally a larger version of those used on smaller pipework and described in Volume 1, Chapter 12. Manufacturers designs differ slightly although the general principles are the same.

Straight and stillson type pipe wrenches are available in lengths from 150 mm (6 in) to 1500 mm (60 in) to deal with pipes from

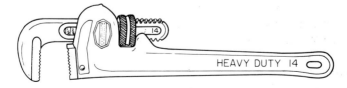

Fig. 17.   Straight pipe wrench

20 mm ($\frac{3}{4}$ in) to 200 mm (8 in) diameter, Fig. 17. The hardened alloy steel jaws are replaceable, handles may be steel, malleable iron or aluminium.

Chain wrenches, Fig. 18, give a very positive grip and can be used in close quarters. The jaws are replaceable and, in some designs, reversible. They are available in similar sizes to the stillsons.

Pipe cuttings tools

Hacksaws may be used on smaller diameter pipe and power hacksaws are used in workshops. Wheel cutters are normally used on larger

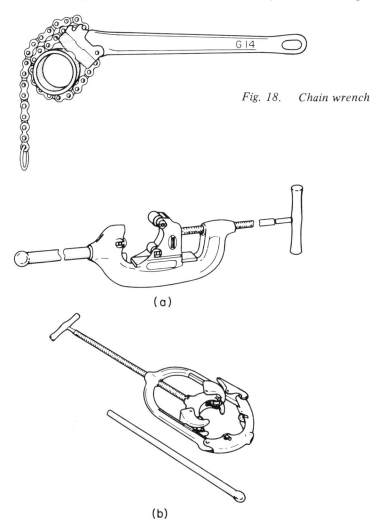

Fig. 18.   Chain wrench

(a)

(b)

Fig. 19.   Pipe cutters: (a) single wheel (b) 4 wheel

diameter pipes. They may have from one to four cutting wheels which are designed for cast iron, ductile iron or steel pipe. Fig.19(a) shows a single wheel cutter for pipes up to 100 mm (4 in) diameter and (b) is a four wheel hinged cutter for pipes up to 300 mm (12 in) diameter. When using cutters it is necessary to remove the burrs formed on the inside and outside of the pipe.

The rollers on the single wheeled type minimise the outer burr but the internal burr must be removed by a reamer. There are several designs and Fig. 20 shows a spiral ratchet pipe reamer for pipes 64 mm ($2\frac{1}{2}$ in) to 100 mm (4 in).

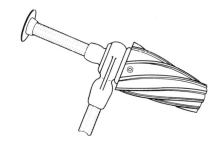

Fig. 20.   Spiral ratchet pipe reamer

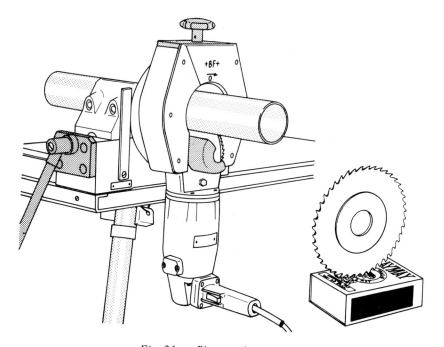

Fig. 21.   Pipe cutting saw

Various types of power operated cutting machines are available. The type shown in Fig. 21 clamps on to the pipe and the machine head, carrying a small circular saw blade, is rotated once around the pipe. The high speed saw gives a square cut without burrs. Bevel cutters may be used in place of the saw to prepare pipe ends for welding.

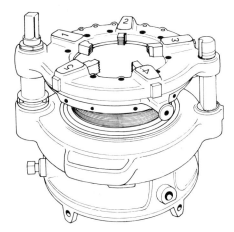

Fig. 22.   Geared receder die stock

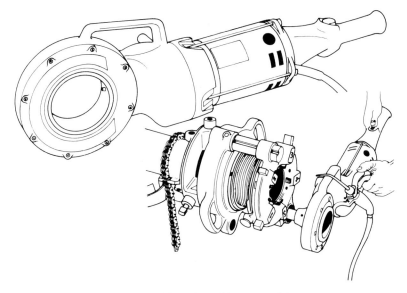

Fig. 23.   Portable power drive

Pipe threading tools

Geared receder dies, Fig. 22, are used for threading pipe from 50 mm (2 in) to 150 mm (6 in). They may be operated manually by a ratchet lever or be connected to a small, portable power drive, Fig. 23. Larger power units, mounted on stands may be fitted with a variety of cutters, die heads or receders.

Many types of screwing machines are available for pipes up to 100 mm (4 in). Above this size the geared receders are used.

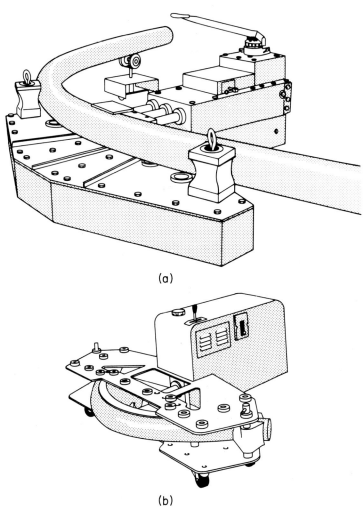

(a)

(b)

Fig. 24.   Hydraulic bending machines: (a) manual (b) power-operated

Pipe bending tools

Bending is usually carried out by hydraulic machines. These are designed to bend pipe up to 150 mm (6 in) and are larger and more robust versions of the smaller models, Fig. 24 (page 27). Most machines are manually operated. Power pumps can substantially reduce the time taken to form a bend and are commonly used for repetitive work.

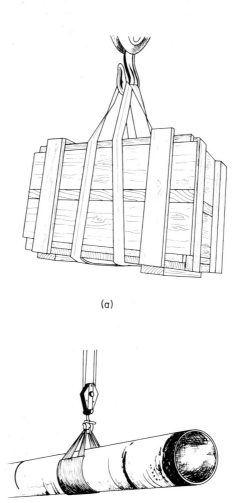

(a)

(b)

Fig. 25.   Slings: (a) nylon webbing (b) pipe sling

Pipe lifting tools

Large diameter pipes are heavy and it requires considerable effort to man-handle them into position, particularly as they are often fitted at high level. In many cases cranes or wire hoists are available on site and pipes and equipment can be moved with the aid of slings. A common type of sling is the woven nylon webbing loop, Fig. 25(a), which is simply passed around the object and looped on to the hook. Special slings are designed to lift large pipes, Fig. 25(b).

When cranes are not available chain hoists may be used either hooked on to roof trusses or supported by shearlegs. The chain hoist has a reduction gear driven by a continuous loop of chain. The gearing drives on to another length of chain with a hook attached for carrying the load, Fig. 26.

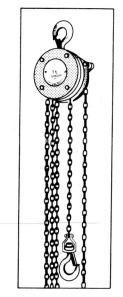

Fig. 26.   *Chain hoist*

It is important that the safe working load (SWL) of the block and slings is not exceeded. If the hoist is attached to any structural members of the building, ensure that these can adequately support the load. All chains, ropes or lifting tackle should be checked before use and examined by a competent person every six months.

§15   Safety Precautions

The safety precautions to be observed when using tools and installing pipework have been described in the previous volumes. They apply equally to large installation work and the following points should particularly be borne in mind:

- do not use a power tool unless you are completely familiar with its operation
- always read the instruction booklet
- make certain that the tool is the right one for the job
  - adequately powered
  - in good condition
  - correctly assembled
- never use a tool with the guard removed
- if a tool does not work, don't tamper with it, repairs should be carried out by an expert
- always unplug the electrical supply before carrying out any adjustments to the tool
- where pipes are stacked on site, secure them against movement with chocks and timbers
- because pipe wrenches are often used at high level and heavy pressures are exerted, the teeth must be clean and sharp, a slip could cause a serious accident
- always wear protective clothing when required; safety helmets, goggles, ear protectors and dust masks are often necessary
- always report any accident whether or not it results in injury
- Get immediate first aid treatment for any injury, even a minor wound

§16 Commissioning

When the installation is completed the following operations should be carried out:

- check that all pipe work is adequately and securely supported
- ensure that test points have been fitted at appropriate places and particularly at the end of pipe runs to enable testing and purging to be carried out satisfactorily
- check that valves have been fitted with handles or hand wheels giving a clear indication whether the valve is open or closed
- before testing for soundness isolate the pipework from any existing installation or meter by plugging or capping off. Valves should not be used as the sole means of isolation
- the joints on underground pipework should not be reinstated until after testing but pipes must be securely anchored and precautions taken to safeguard personnel whilst excavations are open

- pipework must be tested for soundness in accordance with "Soundness testing procedures for non-domestic gas installations, IM/5" (section 17)
- after testing connect to the existing gas supply and test the final connections with leak detection fluid
- ensure that pipework is correctly wrapped or painted and colour coded
- purge the installation in accordance with "Purging procedures for non-domestic gas installations, IM/2"
- any outlets not in immediate use should be capped or plugged
- back fill and reinstate any excavations and wrap or paint any exposed points or connections
- check that drawings of the installation have been fixed by the primary meter and at other appropriate locations
- remove surplus material and clear up the site

§17 Testing for Soundness

The British Gas publication IM/5 was revised in 1979 and retitled "Soundness testing procedure for Industrial and Commercial Installations". The current edition should be consulted before carrying out soundness testing on any large installations.

The procedure applies to non-domestic premises with meters over P4 or D4 size (400 ft^3/h). It does not include small boarding or guest houses. In some special circumstances, for example in cases of extreme temperature variation the procedure cannot be applied. A special test procedure must then be developed of comparable sensitivity.

The procedure should be used on:

- installation of new pipework or appliances
- alteration or replacement of pipework
- installations where there is a known suspected leak

It is intended for pipework downstream of the meter control operating at pressures up to 5 bar (72 lbf/in^2).

On large installations it can be almost impossible to prove the installation sound due to changes in temperature and atmospheric pressure and very small leaks of no significance. So some leakage tolerance must be allowed.

New installations must be tested either with air or an inert gas. On existing installations gas from the service may be used.

Meters, governors and control devices may be tested with the pipework provided that the test pressure does not exceed the maximum

BRITISH GAS

TEST
RECORD
FORM

SOUNDNESS TESTING PROCEDURES FOR EXISTING NON-DOMESTIC GAS INSTALLATIONS

Customer's Name ....   Address ....

....

Job Reference ....   Fitter ....   Supervisor ....

Contractor ....

Working Pressure .... in wg   (If this exceeds 20in wg contact supervisor)

Conditions for the test **must** be steady, therefore the answer to the next two questions must be 'Yes' or 'Not applicable'.
If the answer is 'No' contact the Supervisor.

If part of the Pipework is in the open are the weather conditions stable (steady sun is unacceptable)   Yes/No/Not applicable

Is the Building temperature steady?   Yes/No/Not applicable

Tests which are to include a primary meter, may only be carried out with the permission of the local Region of British Gas.

PROCEDURE

1. Estimate total volume of the installation to be tested from tables 1 and 2. (If this is in excess of 150ft³ consult the supervisor). Pipe Volume .... ft³.
2. Determine the test period from table 3. Test Period .... min.
3. Ensure all appliance isolation valves are shut.
4. Close the inlet valve to section under test.
5. Reduce the pressure by 50% (for example, through a pressure point).
6. Observe the pressure for the test period or 5 minutes, whichever is the longer.
7. If the pressure rise does not exceed 0.2 in wg the valve is satisfactory **for the purpose of this test.**
8. If the valve is unsatisfactory, spade, plug or cap before proceeding. **Valve satisfactory or has the valve been spaded etc.,?** Yes/No.
9. Raise the pressure to the normal working pressure (for example, by opening the inlet valve).
10. Allow the temperature to stabilise for the test period or 5 minutes, whichever is the longer; the pressure during this period should be kept at normal working pressure.
11. Close the inlet valve and observe and record the pressure drop during the test period maintaining the pressors on the **inlet** to the valve within 5% of the pressure in the test section. Governors and non return valves upstream of the pressure gauge should be bypassed during this period.

	Test 1.	Test 2.	Test 3.	Test 4.
Pressure drop (in wg)				
Time (mins/secs)				

12. If the pressure drop does not exceed 0.2 in wg in the test period the installation can be regarded as sound provided that there is no smell of gas anywhere and that pipe joints in unventilated areas (particularly small rooms) have been checked with leak detection fluid. (a) **Is there a smell of gas?** Yes/No. (b) **Pipe joints checked?** Yes/No.
13. If the pressure drop exceeds 0.2in wg attend to the escape and retest recording the results under 11. If unable to trace the leak report the pressure drop to the supervisor. Should the pressure drop in the test period exceed 1 in wg — time the first 1 in wg drop, if not, note the pressure drop in the test period.

Note   For a successful test the answer to 8 and 12b must be 'Yes' and 12a 'No'. If this is not the case contact the Supervisor

SUPERVISING ENGINEERS NOTES

In item 11, governors etc. are temporarily bypassed, for example, using pressure points to avoid 'trapping' pressure. Meter governors should only be bypassed with the agreement of British Gas. If this is not practical the meter governor should be checked separately, for example, using leak detector fluid.

From figure in 11 calculate actual leak rate see notes on the pad of test forms.

	Test 1.	Test 2.	Test 3.	Test 4.
Calculated leak rate ft³/h.				

If leak is below 1ft³/h and the pressure drop exceeds 0.2 in wg.

Is there a smell of gas?   Yes/No

Have unventilated areas and unoccupied areas and ducts been checked with suitable gas detectors (No perceptible movement from 0% LEL or LEL scales) or leak detection fluid?   Yes/No.

In rooms below 1,500 ft³ has each joint been tested with leak detection fluid?   Yes/No

If the leaks have not been traced and the above tests have been satisfactorily carried out the installation may be deemed safe.

If the Leak is above 1ft³/h

Have the above tests been carried out?   Yes/No

Has the customer been informed of the leak rate and the need to report the smell of gas?   Yes/No (this should be confirmed in writing)

If the leaks have not been traced and the above procedures satisfactorily carried out the installation may be deemed safe.

Fig. 27.   Soundness test record form

TABLE 1 (Imperial Units)   PIPE VOLUMES (ft^3)

Length (ft) \ Pipe Size (in)	½	¾	1	1¼	1½	2	3	4	6	8	10	12
1	0.0022	0.004	0.0063	0.012	0.015	0.024	0.055	0.093	0.21	0.37	0.57	0.82
2	0.0044	0.008	0.012	0.024	0.030	0.048	0.11	0.19	0.42	0.74	1.1	1.6
3	0.0066	0.012	0.019	0.036	0.045	0.072	0.17	0.28	0.63	1.1	1.7	2.5
4	0.0088	0.016	0.025	0.048	0.060	0.096	0.22	0.37	0.84	1.5	2.3	3.3
5	0.011	0.020	0.031	0.060	0.075	0.12	0.28	0.47	1.1	1.9	2.9	4.1
6	0.013	0.024	0.038	0.072	0.090	0.14	0.33	0.56	1.3	2.2	3.4	4.9
7	0.015	0.028	0.044	0.084	0.11	0.17	0.39	0.65	1.5	2.6	4.0	5.7
8	0.017	0.032	0.050	0.096	0.12	0.19	0.44	0.74	1.7	3.0	4.6	6.6
9	0.019	0.036	0.057	0.11	0.14	0.22	0.50	0.84	1.9	3.3	5.1	7.4
10	0.022	0.040	0.063	0.12	0.15	0.24	0.55	0.93	2.1	3.7	5.7	8.2
15	0.033	0.060	0.094	0.18	0.23	0.36	0.83	1.4	3.2	5.6	8.6	12
20	0.044	0.080	0.12	0.24	0.30	0.48	1.1	1.9	4.2	7.4	11	16
30	0.066	0.12	0.19	0.36	0.45	0.72	1.7	2.8	6.3	11	17	25
40	0.088	0.16	0.25	0.48	0.60	0.96	2.2	3.7	8.4	15	23	33
50	0.11	0.20	0.31	0.60	0.75	1.2	2.8	4.7	11	19	29	41
60	0.13	0.24	0.38	0.72	0.90	1.4	3.3	5.6	13	22	34	49
70	0.15	0.28	0.44	0.84	1.1	1.7	3.9	6.5	15	26	40	57
80	0.17	0.32	0.50	0.96	1.2	1.9	4.4	7.4	17	30	46	66
90	0.19	0.36	0.57	1.1	1.4	2.2	5.0	8.4	19	33	51	74
100	0.22	0.40	0.63	1.2	1.5	2.4	5.5	9.3	21	37	57	82

Table 2   Meter Volumes

The table gives approximate volumes of diaphragm meter cases; if the meter volume exceeds 1/3 of the installation the case volume should be ascertained by measuring the case.

Meter Size (ft^3) Designation	100 P1 or D1	200 P2 or D2	212 U6	400 P4 or D4	525 U16	700 P7	1200 P12	1800 P18	3000 P30	6000 P60	9000
Volume (ft^3)	0.39	0.78	0.58	1.5	1.0	3.5	4.7	6.7	11	20	38
Meter Size (ft^3) Destination	12000	15000									
Volume (ft^3)	60	68									

Note: (i)   BM meters – measure the case volume (deduct oil volume)

        (ii)   Other meters assume as length of pipe.

Table 3   Test Period

Pipe Volume (ft^3)	0 – 20	20 – 30	30 – 40	For each additional 10ft^3 of part of 10ft^3
Test Period (min)	2	3	4	Add 1 minute

This form is an extract from 'Soundness Testing Procedures for Non-Domestic Gas Installations – British Gas Report No. IM/5.

working pressure of any component. Alternatively, components may be added to the installation after the test has been carried out. They must then be tested at normal working pressure with leak detection fluid.

Testing procedure

The soundness test procedure normally includes:

- estimation of volume of system
- establishment of test pressure
- selection of appropriate pressure gauge
- determination of permitted leakage rate
- determination of duration of test period
- carrying out test

Pads of test record forms are obtainable from British Gas Corporation. The forms have the complete procedure printed on one side and tables giving pipe volume, meter volumes and test periods on the other. A copy of a form for new installations is shown in Fig. 27. The pads are obtainable in either S.I. or Imperial units and designed for new or existing installations.

Volume of system

This requires a survey of the whole installation. Short lengths of small diameter pipes may be ignored. Volumes of pipes and meter may be taken from the test form. Valves and fittings, RD, turbine or orifice plate meters should be taken as part of the pipe.

Test pressure

On new installations the test pressure should be:

50 mbar (20 in w.g.)
or $1\frac{1}{2}$ times the working pressure
or   the maximum pressure likely to occur under fault conditions
whichever is the greatest.

On existing installations:

not less than the normal working pressure
On installations which will be tested at pressures above 3 bar (45 lbf/in^2) consideration should be given to applying a hydraulic test first.

Pressure gauges

The following types are normally used:

- water gauge—0 to 50 mbar (0 to 20 in w.g.)
- high s.g. fluid gauge—50 to 120 mbar (20 to 48 in w.g.)
- mercury gauge—50 to 700 mbar (20 in w.g. to 10 lbf/in²)
- dead weight oil column—above 700 mbar

Permitted leakage rate

On new installations and extensions this is

1.4 dm³/h (0.05 ft³/h)

On existing installations the rate varies with the location of the pipe.

In hazardous areas it is not considered safe to rely entirely on the pressure test and all joints in the area should be tested with leak detection fluid or a gas leakage detector.

In occupied areas a leak which cannot be smelled is considered insignificant. If the threshold of smell is 0.05% gas in air then at an average air change the leakage rate is 0.75 dm³/h per cubic metre of space (0.00075 ft³/h per cubic foot of space). In no case may it be greater than 30 dm³/h.

Test period

Table 4 shows the leakage per cubic metre of pipe and the equivalent pressure drop. It also gives the maximum test period. In some circumstances it may be necessary to exceed these times but this may require corrections to be made for changes in barometric pressure.

§18 TABLE 4

Type of Gauge	Minimum Pressure Drop Reading mbar	Leakage per Cubic Metre of Pipe dm³	Maximum Test Period min
Water gauge	0.5 *0.25	0.5 0.25	15
High s.g. fluid gauge	1.0 *0.5	1.0 0.5	30
Mercury gauge	3.4 *1.7	3.3 1.7	60
Dead weight oil column	0.7 *0.35	0.68 0.34	30

*The upper figure is the normal minimum gauge reading. The lower figure is half the normal and is regarded as "no perceptible movement" on the gauge.

The duration of the test period for a new installation may be calculated as follows:

$$\text{Test period (min)} = \begin{array}{c}\text{Minimum detectable}\\\text{leak per m}^3\text{ of pipe}\\\text{(dm}^3\text{)}\end{array} \times \begin{array}{c}\text{Installation}\\\text{volume}\\\text{(m}^3\text{)}\end{array} \div \begin{array}{c}\text{Permitted}\\\text{leak rate}\\\text{(dm}^3/\text{h)}\end{array}$$

Example:

Gauge used—Water

Minimum detectable—0.25 dm^3
leak/m^3 pipe

Permitted leak—1.4 dm^3/h $\left(\text{in minutes} = \dfrac{1.4}{60}\right)$

Installation volume—0.35 m^3 (say)

$$\text{Test period} = 0.25 \times 0.35 \div \frac{1.4}{60} \text{ min}$$

$$= \frac{0.25 \times 0.35 \times 60}{1.4}$$

$$= \textbf{3.75 min}$$

Rounded up to nearest minute

$$= \textbf{4 min}$$

Carrying out test

When the previous operations have been completed, the test should be carried out as follows:

- seal off the section to be tested, cap or plug the isolating valves and then reopen so that the test is against the sealed ends
- raise the installation pressure to that required for the test
- allow the temperature to stabilise for at least 15 minutes while maintaining the test pressure
- isolate the pressure source and observe the gauge for the test period
- if there is any perceptible movement of the gauge, repair the leak and retest
- after a satisfactory test, any joints in potentially hazardous areas should be checked with leak detection fluid and any leaks repaired
- unseal the tested section and test the isolating joints with leak detection fluid; deal with any leaks
- purge the installation and put into service

when the installation is in use check all joints in potentially hazardous areas with a gas detector and repeat the test after 3 to 4 days (there should be no perceptible movement for 0% LEL on the LEL scale)

On existing installations, governors or non-return valves may require to be bypassed by flexible tubing between the pressure points.

Isolating valves must first be checked to ensure that they are not letting by.

An alternative method of test is to use a test meter or a calibrated pump to supply gas from an independent source to maintain the test pressure constant. The volume of gas supplied is equal to the leakage rate. This method is particularly useful where the installation volume is not known.

§19   Purging

The Gas Safety Regulations require that after testing an installation for soundness it must immediately be purged. The procedure to be followed is described in B.G.C. report IM/2. "Purging Procedures for Non-Domestic Gas Installations".

Purging is required after the following operations:

- installing, exchanging or removing meters
- installing, altering or removing pipework
- installing, servicing or removing control devices
- after a valve closure resulting in a loss of pressure and possible admission of air
- after incorrect use of equipment using air or another gas

Report IM/2 defines purging as:

- displacing air or inert gas by fuel gas
- displacing fuel gas by air or inert gas
- displacing one fuel gas by another

The procedures are applicable to all sizes of pipework from downstream of the meter inlet valve to the end of the supply or the appliance inlet, irrespective of the working pressure.

Basic requirements

A soundness test must always be made before purging and, as with testing, purging should not take place unless the person in charge has a full knowledge of the pipework. All plans should be checked for accuracy.

In all cases, except for very simple purging operations, a written procedure for the particular installation should be prepared and followed.

At a distance up to 3 m (10 ft) from the vent point or test burner no naked lights should be allowed except when a vent gas is being ignited. "No smoking" notices should be displayed where necessary.

Warning notices should be displayed on valves to or from the section being purged. They should state "Purging in progress, do not open this valve". Care should also be taken to ensure that no other work is taking place on the pipework being purged. This can be controlled by the issue of "Permits to work".

Ensure that there is adequate ventilation at vent points and in confined spaces and basements where accumulations of inert gases may occur.

At all times precautions should be taken to prevent air or inert gas from entering British Gas mains and no purge should be left incomplete.

Purge volumes

The purge volume for a diaphragm or BM meter is 5 revolutions of the meter. On other meters the purge volume is $1\frac{1}{2}$ times the volume of the length of pipe equal to the flange to flange length of the meter.

The purge volume for pipework is $1\frac{1}{2}$ times the volume of the pipework. Table 5 gives the approximate volumes of 3 m (10 ft) lengths of various diameter pipes.

TABLE 5 Approximate Volumes of 3 m (10 ft) Lengths of Pipe

Diameter		Volume	
mm	(in)	dm^3	(ft^3)
15	$\frac{1}{2}$	0.61	0.022
20	$\frac{3}{4}$	1.1	0.040
25	1	1.8	0.063
40	$1\frac{1}{2}$	4.1	0.15
50	2	6.6	0.24
65	$2\frac{1}{2}$	11	0.40
80	3	15	0.55
100	4	26	0.93
150	6	60	2.1
200	8	100	3.7
250	10	160	5.7
300	12	220	8.2

The purge volumes given should only be used as a guide. The only way to verify that the purge is complete is to test the vent gas. This is normally done by a gas detector. Allowing the gas to burn at the test burner may be used but not when displacing air with fuel gas.

The purge volume may be read on the existing meter or on a flow meter fitted on the vent line.

Purging methods

Details of the purging procedures appropriate to the various purging methods are given in Report IM/2.

The types of purge are:

- direct purge
- inert purge — complete displacement
                     — slug purge
- fuel gas to fuel gas purge

Direct purge

This is the displacement of air or an air/gas mixture with fuel gas. It is the method most commonly used to put pipework into service.

Inert purge

Complete displacement is the replacement of fuel gas or air by an inert gas. Slug purging is the formation of a barrier of inert gas between fuel gas and air. Inert purging is used for pipes of 80 to 100 mm (3 to 4 in) diameter in runs of 100 m (300 ft) or over.

When putting pipework into service the inert purge should be followed by a direct purge. Figure 28 shows the pipe layout for an inert purge.

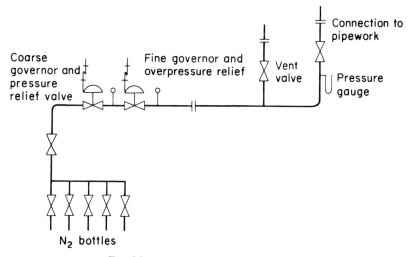

Fig. 28.   Pipe layout for inert purge

Air purge

This is the displacement of gas or air/gas mixture with air. It is, in effect, a direct purge in reverse and is used when taking pipework out of service.

Fuel gas purge

This is similar to a direct purge. The gas may be burnt off at an appliance or on a special vent gas burner.

Direct purging procedure

Preparations should be made in accordance with the basic requirements and the system volume calculated.

Vent points, Fig. 29, should be installed at the extreme end of every branch of the pipework to be purged.

After the soundness test has proved satisfactory, purging should

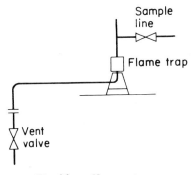

Fig. 29.   Vent point

begin by opening the vent points and by admitting the fuel gas. Any number of vent points may be operated simultaneously, provided that each is supervised.

The gas should be admitted as quickly as possible and branches should be purged in reducing order of pipe size. Ensure that ring mains or interconnected branches are completely purged.

When the purge volume of gas has been admitted, test the vent gas at the vent points and close and seal each point when the test is satisfactory. After sealing, check each disturbed joint with leak detection fluid.

Venting

Where the purge does not release more than 25 dm^3 (1 ft^3) of natural gas or 0.1% of the room volume, whichever is the smaller, the vent gas may be released in the room without a flame arrestor. This is provided that there is good ventilation and no source of ignition nearer than 3 m (10 ft).

Otherwise the gas from the purge should be vented into the open air at least 2.5 m (8 ft) above ground level and 3 m (10 ft) from any source of ignition.

Ensure that the vent gas does not drift into buildings or accumulate in ducts, pits or confined spaces.

The vent point may be connected to the purge outlet by reinforced flexible hose. This must be firmly secured to both points and, if externally armoured, it must be earthed to prevent sparking.

The vent line should be fitted with a flame arrestor. This may be at the vent outlet, if the gas is not to be ignited. If the gas is to be ignited, the flame trap must be located between 0.5 m and 1 m (18 in and 3 ft) from the vent outlet.

Vent gas may only be ignited on the following purges:

- fuel gas to fuel gas
- inert gas to fuel gas
- fuel gas to inert gas

Fuel gas must never be ignited at the flame arrestor.

Vent gas testing

IM/2 gives safe levels of oxygen and combustibles for the various purge methods. When air is being displaced by natural gas in a direct purge the oxygen level should be less than 4% and the combustible level greater than 90%.

The gas detector should be used in accordance with the manu-facturer's instructions. It should be flame proof. The sampling line should be checked for leaks prior to testing and the calibration of the detector checked before and after a purge.

A combustion test should not normally be carried out on an air to gas purge. However, on very small installations with a pipework volume of less than 25 dm^3 (1 ft^3) a test burner may be used. This must be a bunsen burner and, to prevent light back the injector jet diameter must be less than 1 mm and the pressure must be above 3 mbar (1.2 in w.g.).

With air to inert purging the oxygen level should be less than 4%, whilst the inert to fuel gas purge should result in a combustible level greater than 90%.

CHAPTER 2

Non-Domestic Gas Meters

Chapter 2 is based on an original draft by Mr. R. H. Wharton

§20 Introduction

The basic principles of the common types of gas meter were outlined in Volume 1, Chapter 7. This showed meters to be measuring devices which:

- recorded the total quantity of gas which has passed through them
- indicated the rate of flow at any moment of time

So the information provided by a meter could be used as a:

- basis for charging gas used
- means of measuring the gas rate of a burner

Under the Gas Act, 1972, meters used as the basis for a charge must be stamped or "Badged", except where a customer is supplied under the terms of a special contract. The stamping is carried out by a meter examiner when a meter meets the requirements of the Gas (Meter) Regulations, 1974, for accuracy and pressure loss.

Badged meters must be accurate to within ± 2% and permitted maximum pressure drops **for diaphragm meters only** are as follows:

Diaphragm meters, designed for pressures not exceeding 1 bar.

Badged Rating	Pressure Drop
not above 565 ft³/h (16 m³/h)	up to 2 mbar
above 565 ft³/h but not above 2296 ft³/h (65 m³/h)	up to 3 mbar
above 2296 ft³/h	up to 4 mbar

Rotary displacement and turbine meters have no statutory maximum permissible pressure drop.

British standard requirements for meters have been modified to permit the exchange of meters with other EEC countries. In addition, international regulations for most types of gas meter are prepared by the International Organisation for Legal Metrology (OMIL).

42

BS 4161 covers diaphragm, rotary displacement and turbine meters and also mechanical volume correctors.

§21 Types of Meters

Meters may be classified as:

- inferential
- positive displacement

In the inferential types, the quantity of gas flowing may be inferred from any of the following measurements:

- speed rotation of a turbine
- pressure loss across an orifice
- difference between static and kinetic pressures
- change of temperature in a wire
- height of a rotating float in a tapered tube

In positive displacement meters, a definite volume of gas is measured by displacement through one of the following:

- bellows or diaphragms
- compartments submerged in a liquid
- spaces between impellers or vanes

At one time, only the positive displacement meters were badged and all inferential meters were not. This is no longer the case and perhaps it would now be better to categorise meters as either badged or non-badged.

BADGED METERS

§22 Diaphragm meters

Tin-case meters

The tin-case meter was available in sizes up to 50 000 ft³/h (1415 m³/h) but it is now no longer made. In recent years it has rarely been used above 6000 ft³/h (170 m³/h).

It is being superseded by the steel-case, unit construction meter, for the smaller sizes and rotary and displacement or turbine meters for the larger.

Steel-case meters

The "U" series of steel-case meters is shown in Table 1.

Table 1   Range of Unit Construction Meters

Designation	Badged Rating		Connections		
	m³/h	ft³h	Type	Nominal Bore	Centres
U6	6	212	Threaded	25 mm (1 in)	152 mm (6 in)
U16	16	565	Threaded	32 mm (1¼ in)	152 mm (6 in)
U25	25	883	Threaded	50 mm (2 in)	250 mm (10 in)
U40	40	1412	Threaded	50 mm (2 in)	280 mm (11 in)
U65	65	2295	Flanged	65 mm (2½ in)	335 mm (13 in)
U100	100	3530	Flanged	80 mm (3 in)	430 mm (17 in)
U160	160	5650	Flanged	100 mm (4 in)	430 mm (17 in)

Fig. 1.   'U' series meter

The U6 meter, shown in Fig. 1, is described and illustrated in Volume 1, Chapter 7, (142). It was developed from the D07 meter for domestic use and its success prompted the redesign of the larger meters.

Because they were designed to operate at lower pressure drops, the tin-case meters may be allowed to operate at an overload rate.

'U' series meters, however, must not be subjected to loads in excess of their badged ratings. Overloading the meter would cause it to operate at too high a speed, resulting in excessive pressure drop. It would also invalidate the maker's guarantee.

A comparison of the essential features of tin-case and steel-case meters is shown in Table 2.

TABLE 2 Comparison of Tin-Case and Steel-Case Meters

Feature	Tin-case	Steel-case
Size of case for same rating	Larger type of meter	Smaller and more compact
Normal operating pressures	50 mbar	200 mbar
Permissible pressure drop	Low (1.25 to 2.5 mbar) (under old regulations)	Higher (2 or 4 mbar)
Maximum load	Some overload permissible	As badged rating
Strength and fire resistance	Poor. Soft soldered seams can pull apart or melt easily	Good. Withstands the $\frac{1}{2}$ hour fire test
Installation	Flexible connections required. Meter must be supported	Rigid pipework may be used meter has inbuilt meter bar
Connections	Up to 700 ft³/h, upward facing, screwed BS 746 1200 ft³/h and above, side facing, socket or flange	All sizes, upward facing. Up to U40, screwed BS 746. Above U40, metric flanges

Tin-case meters always require protecting from the weather and are often located in special housing. Generally steel-case meters are similarly protected although they may sometimes be fitted outdoors.

High pressure aluminium-case meters

High pressure aluminium case meters are basically similar to unit construction meters. They are enclosed in die-cast aluminium cases which can withstand much higher working pressures.

Sizes range from 175 ft³/h (5 m³/h) to 5000 ft³/h (142 m³/h). Working pressures are from 350 mbar to 7 mbar respectively.

This type of meter is used for high pressure loads where rotary meters are unsuitable. That is, when:

· the load is subject to rapid on-off cycling
· there are low rates of flow for a large proportion of the time

Figure 2 shows a range of aluminium case meters with a sectional view at b.) Aluminium-case meters may be fitted outdoors.

§25 Rotary Displacement Meters

Roots-type meter

This was described in Volume 1, Chapter 7 (137). It measures gas by trapping fixed volumes between two intermeshing impellers and the

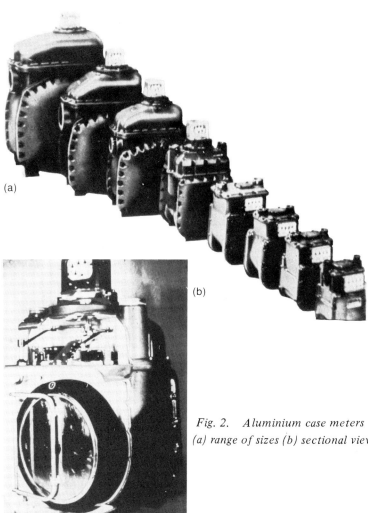

(a)

(b)

Fig. 2.   Aluminium case meters
(a) range of sizes (b) sectional view

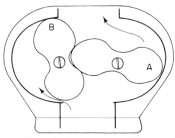

Fig. 3.   Operation of Roots-type meter (Holmes)

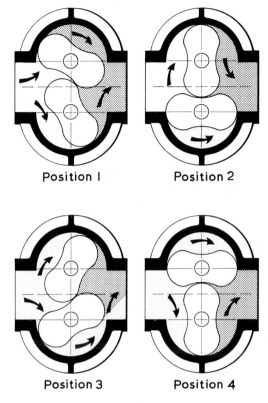

Position 1   Position 2

Position 3   Position 4

Fig. 4.   Sectional view of Roots meter (Dresser)

Fig. 5.   Roots meter installation (Holmes)

casing, Fig. 3. The impellers are geared together so that they rotate in opposite directions. A sectional view of a typical meter is shown in Fig. 4 and a meter installation in Fig. 5.

Roots type meters are available in a range of sizes from $1500 \, \text{ft}^3/\text{h}$ $(43 \, \text{m}^3/\text{h})$ to $320\,000 \, \text{ft}^3/\text{h}$ $(9063 \, \text{m}^3/\text{h})$.

Rotary vane meters

These are also known as "vane and gate" or "rotary vane meters.

They have an annular measuring chamber formed between the casing and a central, stationary cylinder. Vanes rotate around the chamber, trapping fixed volumes of gas between them and passing the gas from inlet to outlet. A rotor, geared to the vanes, forms a gate which allows the vanes to return to the inlet but prevents gas from bypassing the measuring chamber. Depending upon the manufacturer, the vanes and the gate may rotate in the same direction, or in opposite directions. The operation of the two types is shown in Fig. 6.

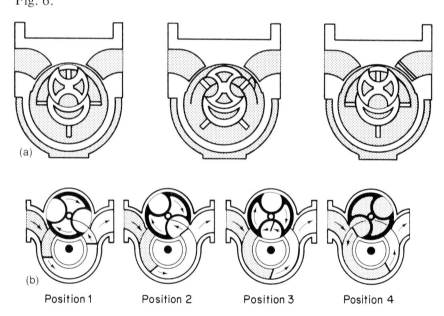

| Position 1 | Position 2 | Position 3 | Position 4 |

Fig. 6.   Operation of rotary vane meters (a) I.G.A. (b) Rockwell

Rotary vane meters are available in a range of sizes from $3000 \, \text{ft}^3/\text{h}$ $(85 \, \text{m}^3/\text{h})$ to $38\,000 \, \text{ft}^3/\text{h}$ $(1076 \, \text{m}^3/\text{h})$.

Figure 7 shows sectional views of the two types of meter and the actual meters are shown in Fig. 8.

Fig. 7. *Sectional view of rotary vane meters (a) I.G.A. CVM (b) Rockwell Rotoseal*

General features

Although there is no statutory limit to the pressure drop through rotary displacement meters, too high a pressure drop would be

(a)

Fig. 8   Rotary vane meters
(a) I.G.A. CVM (b) Rockwell Rotoseal

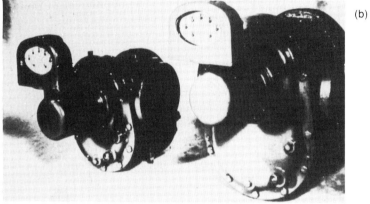

(b)

unacceptable on low pressure installations. The pressure drop increases with the rate of flow to a maximum of about 5 mbar at the badged rating. Where this is too high, the meter should be sized either on the basis of an acceptable pressure drop or on the old badged rating. This allowed a maximum pressure drop of 2.5 mbar.

As an example, a meter which has a pressure drop of 5 mbar at its badged rating of 20 000 ft^3/h (566 m^3/h) would, under the old conditions, have been badged at 10 500 ft^3/h (297 m^3/h) at a pressure drop of 2 mbar.

The inertia of the rotors or impellers can cause pressure fluctuations as the flow rate changes suddenly. On starting up this will result in a momentary pressure drop. Conversely, when the rate is suddenly reduced, a surge of higher pressure will occur. Adequately sized pipework on the meter outlet will reduce the effect of the fluctuations. The effect may also be minimised by using multiple small governors, having quick response, rather than one large, slower acting control.

Accuracies always within ± 2% and generally within 1% may be obtained when measuring flows over a turn-down of about 30:1 and meters must be selected for their accuracy at the minimum and maximum flow rates required (See Gas (Meter) Regulations 1974, para. 4 (1) (c).)

All rotary displacement meters can operate at pressures up to 2 bar and some are suitable for up to 100 bar. At high pressures it is necessary to fit straight lengths of pipe of 4 diameters on the inlet and 2 diameters on the outlet. Straightening vanes are not required.

If the meter stops rotating, it will shut off the gas supply. So, if the supply must be maintained, it is necessary to fit a bypass or to have two meters in parallel streams. This ensures continuity of supply and allows a meter to be shut down for maintenance.

These meters are capable of driving ancillary equipment, for example, a mechanical or electrical volume corrector. They are more robust than diaphragm meters and may be fitted outside without any special protection from the weather. In sizes up to 11 000 ft^3/h (312 m^3/h) they may be supported by the pipework or "line-mounted". Above this size they must rest on a plinth or base, as in Fig. 5.

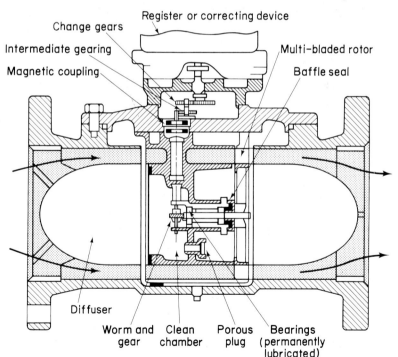

Fig. 9.   Principle of the turbine meter (I.G.A.)

§26   Turbine Meters

These are similar to vane anemometers and measure the speed at which gas flows through an annular gas-way of known cross-sectional area.

The gas-way is formed by a diffuser cone situated centrally in the body and a light-weight air foil turbine is placed so that the blades rotate around the annulus, Fig. 9. The speed of the rotation of the turbine is proportional to the velocity of the gas through the gas-way and also proportional to the volume of gas flowing.

The drive from the turbine spindle is transferred to the index mechanism by a magnetic coupling, so eliminating the need for a stuffing box.

Because the gas is made to flow through a narrow annulus, its velocity is increased. This provides the necessary energy to drive the mechanism and gives more accurate registration.

As the turbine rotates freely, it has a tendency to overrun when flow stops. So the meter is best suited to situations where the gas flow is generally constant. Some models have an aerodynamic brake to minimise overrun.

Meters are available in a range of sizes from 8800 ft³/h (249 m³/h) to 230 000 ft³/h (6514 m³/h). Figure 10 shows a sectional view of a turbine meter and other types are shown in Fig. 11.

Fig. 10.   Sectional view of turbine meter (P.C.C.)

In common with the rotary displacement meters there is no statutory maximum permissible pressure drop through turbine meters. Pressure drop increases with the rate of flow up to almost 9 mbar at the badged rating.

This is too high for low pressure installations and meters should be sized on the basis of an acceptable pressure drop. For example, a meter might have a pressure drop of 4 mbar at 140 000 ft³/h (3965 m³/h) but a pressure drop of only 2 mbar at 100 000 ft³/h (2832 m³/h). Manufacturers will give details of the pressure drop

(a)

(b)

Fig. 11.   Turbine meters (a) range of I.G.A. (b) Rockwell

through their meters at various rates of flow. A graph showing the relationship for turbine and rotary displacement meters is given in Fig. 12.

Because of the tendency to overrun, the turbine is not suitable for loads where gas only flows for short periods during the day, or where the flow cycles rapidly on and off.

At low pressure, accuracy is maintained over a turn-down of 15:1. This range increases as the pressure is increased and meters must be selected to have an acceptable accuracy at the minimum and maximum flow rates required.

All turbine meters can operate at pressures up to 7 bar and some are suitable for up to 100 bar. Some meters have integral straightening vanes while others have them supplied separately by the manufac-

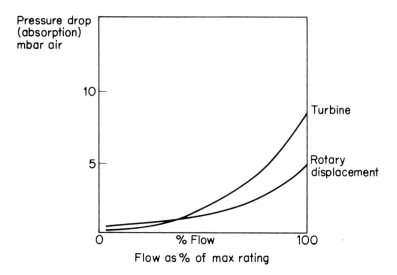

Fig. 12. Graph of pressure drop with flow for turbine and RD meters

turers. These take the form of a pipe section of 4 diameters in length containing a tube bundle at one end. The straightening device is fitted to the inlet of the meter with the tube bundle farthest from the meter. The device is supplied with each badged meter.

If the meter stops rotating for any reason, gas can continue to flow. So a bypass is not strictly necessary for continuity of supply. It is, however, necessary to allow the meter to be shut down for maintenance or exchange.

These meters are capable of driving ancillary equipment through either a mechanical linkage or an electrical pulse. There are, however, limitations to the type of instrument which may be driven mechanically without affecting the registration and both manufacturers and the Department of Energy instructions must be followed.

Turbine meters may be fitted outside without special protection and may generally be lined-mounted. Where meters are fitted with ancillary equipment it is advisable to mount them on a support.

NON-BADGED INFERENTIAL METERS

§27 Turbine Meter Anemometer

Essentially similar in design to the rotary meter described in Volume 1, Chapter 7 (151), this meter is used as a non-domestic secondary meter, Fig. 13.

The rotor has flat, aluminium blades and is usually mounted vertically in a central measuring unit which may easily be withdrawn for

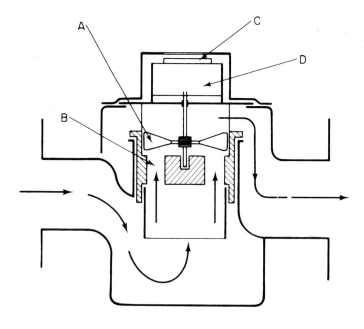

A. Flat bladed turbine
B. Guide ports
C. Index
D. Integrating mechanism

Fig. 13.   Vane or turbine anemometer

servicing or repair. Gas is directed on to the rotor blades through guide ports in the base of the unit. A drive from the rotor spindle operates an integrating mechanism and the gas consumption is shown on the index on the top of the meter.

These meters are available in a range of sizes up to $7000 \, \text{ft}^3/\text{h}$ ($200 \, \text{m}^3/\text{h}$). Although not designed to be badged, their accuracy may be within $\pm 2\%$ over a 10:1 turndown. They can operate at pressures up to 1.7 bar ($25 \, \text{lbf/in}^2$) and must be fitted level in horizontal pipework.

INFERENTIAL METERS

§28   Orifice Meters

Meters which indicate gas flow by measuring differential pressures were introduced in Volume 1, Chapter 5 (120) and Chapter 7 (136). Of these, the orifice meter is the most commonly used, mainly because of its simplicity and low overall cost. It is capable of operating at high pressures and is frequently used for measuring the flow of gas through transmission networks but results in a significant pressure drop across the meter. Although not a stamped meter, it can be used for the sale of gas on a contract basis.

Orifice plates are made as thin as possible and bored out to produce a sharp, square-edged, circular orifice. They are usually fitted between flanges with the orifice concentric with the pipe bore, Fig. 14. When a thicker plate is used to maintain rigidity, the orifice is chamfered on the outlet edge. The material used for the plate is a non-corrodible metal such as stainless steel or monel metal.

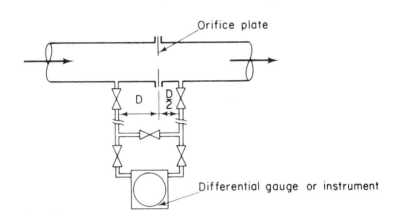

Fig. 14. Orifice meter

Accuracy of measurement depends on the condition of the plate and regular maintenance is necessary to ensure that it does not become dirty, distorted or eroded. It is also necessary to have straight, unobstructed pipework for about 20 diameters upstream and 15 diameters downstream of the meter. Orifice meters are unsuitable for situations where the flow rate varies considerably or pulsates. However, different flow rates can be metered by simply changing the orifice and/or the differential instrument. Maximum flow can be increased by up to 16 times in the same diameter pipe. Accuracy of ± 2% can be maintained over a turndown of about 4:1.

Meter failure has no effect on continuity of supply and a bypass is not essential. It is, however, useful for pressurising downstream pipework and for maintenance.

The design of orifice meters is specified in BS 1042, Part 1.

29 Other Instruments

These brief notes are included for information if required.

Venturi meter (Fig. 15)

This has the advantage of creating a lower pressure loss than the orifice plate and is used on low pressure systems. The flow rate through the meter can be determined by a differential pressure gauge or by a positive displacement meter, used as a "shunt" meter and measuring about 1% of the total gas flow.

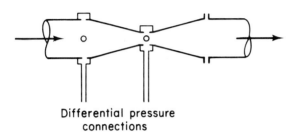

Differential pressure
connections

Fig. 15.   Venturi meter

Pitot tube (Fig. 16)

This is seldom used as a permanent meter installation. It has the advantage of portability and is generally used for checking flows on mains networks. It requires very careful use to obtain accurate readings. The point of the tube must be in line with the flow of gas and positioned at the point of mean velocity.

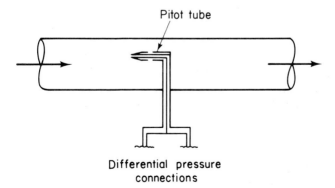

Pitot tube

Differential pressure
connections

Fig. 16.   Pitot tube

Insertion meters

As an alternative to the pitot tube, a small turbine wheel on the end of a rod may be inserted into the pipe. The rotation of the turbine is conveyed to an external indicating device and the meter can measure the average velocity of the gas flow. It is not highly accurate.

Sonic meter

A further development is the use of sound waves to measure the speed of gas flow. This "sonic meter" can measure large flows of gas at high pressures, over a wide range of temperatures.

The instrument measures the time taken for a sound wave to travel between two probes inserted just inside the pipe. The probes are on opposite sides of the pipe and at some distance from each other. Measurements are taken both with, and against, the gas flow. The times are fed into a computer which calculates and displays the rate of flow. By taking measurements at fixed time intervals, the total volume is also computed.

The equipment is portable and requires only valved branches into which the probes may be inserted. The meter has an accuracy of about ± 0.75% at flows above 1.5 m/s (5 ft/s).

Hot wire anemometer (Fig. 17)

In one form, the gas flow passes over an electric heater, A, situated between two resistance thermometers B and C. The heater is controlled by the thermometers and raises the temperature of the gas by about 1°C. The amount of energy required to do this is measured by a watt hour meter which is calibrated to give a direct reading of gas volume.

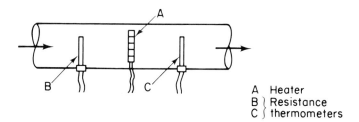

A   Heater
B ⎱ Resistance
C ⎰ thermometers

Fig. 17.   Hot wire anemometer

Rotameter (Fig. 18)

This instrument is commonly used in laboratories and on appliance testing rigs in workshops. It consists of a float in a transparent, tapered, vertical tube. Gas flows through the annular gap around the float and so creates a pressure difference. This supports the float which has small slots in its outer edge so that it rotates as it is suspended on the upward flow of gas. The height of the float in the tube is proportional to the rate of flow of the gas, which may be read on a scale at the side of the tube.

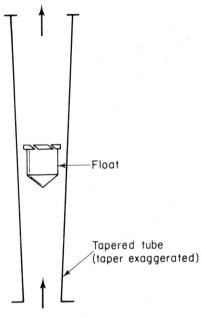

Fig. 18.   Rotameter

§30   Installation of Meters

The installation of gas meters used as a basis for charging is covered by:

- a BS Code of Practice for low pressure metering, with pressures at the meter control not above 75 mbar (30 in w.g.)
- I.G.E./GM/1, Installation of High Pressure Gas Meters for inlet pressures above 75 mbar

The Gas Safety Regulations make some of the recommendations of the above codes obligatory. They should be studied in association with the recommendations.

The installation of non-domestic low pressure diaphragm meters is, in many respects, similar to the installation of domestic meters. The requirements given in Volume 2, Chapter 4 are applicable in many cases.

Rotary displacement and turbine meters must be installed in accordance with manufacturer's instructions.

Points of particular importance are as follows:

- filtering
- levelling
- cleaning and oiling
- pressurising

Filtering

Gas borne solids, in the form of dust or rust particles can effect the operation of meters.

Diaphragm meters are not generally affected but a coarse filter is recommended.

Orifice plates may be eroded by dust and deposits can affect registration. A fine filter is required, capable of stopping particles larger than the 2 micron.

Rotary displacement meters require filters to stop particles larger than 250 micron. With some meters filtration down to 50 micron is necessary. Where filters are fitted upstream of the governors, it is advisable to fit a "top-hat" filter on the inlet of the meter. This should face the gas flow, as shown in Fig. 19.

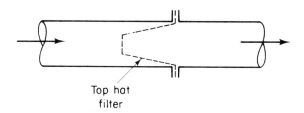

Top hat
filter

Fig. 19.   Top-hat filter

Turbine meters should be protected with filters capable of removing particles larger than 250 micron to prevent damage to the blades and erosion.

Filters should have ease of access for cleaning or exchanging. Trapped dust should not be allowed to fall back into the system.

Meters can also be damaged by slugs of liquid and suitable receivers and drain points may be necessary.

Levelling

Most meters should be fitted to reasonably horizontal connections. Rotary displacement meters must be levelled accurately to tolerances specified by the manufacturer. Levelling is carried out along the axis of the roto to an accuracy of about 1 in 192 (1/16 in. in 1 ft). This must be done with a precision spirit level.

Turbine meters are less critical and one type may be installed at any angle.

Cleaning

Rotary displacement meters have their internal parts treated with a protective coating to prevent corrosion in storage. They should have this coating removed from the rotors and measuring chambers before installation by means of a non-flammable solvent.

Oiling

Diaphragm meters which are fitted with external lubrication points should be oiled in accordance with the manufacturer's instructions.

The oil used is a vegetable-based-oil unsuitable for use in other types of meter.

Most rotary displacement meters have gear boxes at either end which should be filled to the required level with a light, mineral-based oil to BS 4231, Grade 10, before putting the meter into operation. The oil should be changed after one month's operation and checked annually.

Turbine meters may require their bearings lubricated at intervals. This is normally carried out when the meter is given periodic maintenance.

Pressurising

Most types of meter may be damaged if there is an excessive pressure difference across the meter. On diaphragm meters this can puncture diaphragms and strain linkages. Where pressures are high, pipe work downstream of the meter must be pressurised through a bypass before the meter outlet and inlet valves are opened, in that order.

Rotary displacement and turbine meters should not be pressurised or depressurised at a rate of more than 350 mbar/s (5 lbf/in^2/s). This is most easily carried out by means of a bypass. If no bypass is fitted, pressurise the meter by opening the inlet valve and then pressurise the downstream pipe work by means of the outlet valve.

The same rate of pressurisation should be used for orifice meters to avoid deforming the plate. Pressures must be equalised across the pressure differential instrument and its isolating valves closed.

With all types of meter, the same care is necessary when depressurising the system.

§31 Location of Meters

Low pressure meters

Inside a building a primary meter should be located:

- as near as possible to the point of entry of the service pipe
- as near as practicable to the site boundary
- with adequate access for repair, maintenance or exchange
- to permit the installation or servicing of control devices or ancillary equipment
- in a well-ventilated position
- protected from any corrosive material or atmosphere
- remote from any heat source
- to avoid causing an obstruction
- where there is no risk of damage

Outdoor locations should be chosen to give free access for repair, maintenance or exchange. The site must be well drained and not subject to flooding. The installation may need to be fenced and protected from vehicles and vandalism. Meters should not normally be exposed to temperatures below −5°C.

Where the meter is located in a purpose-built house the following points should be noted:

- planning permission may be required and local bylaws must be observed
- the materials must be—weather and waterproof
                     —fireproof to BS 476
                     —strong and durable
- there should be a substantial foundation
- the housing must allow access for repair, maintenance or exchange of the meter or any item of ancillary equipment
- access must be provided to permit the use of lifting gear, where required
- the housing should be fitted with outward opening doors and be adequately ventilated
- any electrical equipment installed must be to the required safety standards
- the housing should not be occupied by personnel or used to store materials

High pressure meters

On non-domestic premises these should be located as close to the boundary fence as possible, to reduce the length of high pressure pipework to a minimum, particularly indoors. The installation should preferably be outside and away from buildings or in a separate meter housing.

If this is not possible, a special room should be constructed within the building to house the installation. It should have external, outward opening doors, gas tight internal walls and adequate ventilation to outside air.

If the meter is separate from the pressure governors and in the same building as the industrial process, the foregoing precautions may not be necessary.

In addition to the requirements for low pressure meters the following points should be observed.

- external sites should be clear of high-tension electric cables and large trees
- care must be taken to avoid noise from the installation causing a nuisance
- any outdoor installation should be fenced with access for repair, maintenance or fire-fighting
- any housing should have two sets of outward opening double doors, preferably on opposite sides of the building
- an explosion relief may be desirable, this could be incorporated in the roof design
- relief valves have vent pipes terminating outside the building
- the total ventilation area should be not less than 2% of the floor area and equally dispersed at the top and bottom of the walls

32   Bypassing Meters

A full-flow bypass may be required where complete continuity of supply is essential and failure of the meter would affect the flow.

Similarly, where the full rate of flow must be maintained during maintenance operations, a bypass or temporary rider is required.

Where full-flow is not required, a low-flow bypass may be fitted to assist commissioning. It is advisable always to fit a bypass to a high pressure meter to facilitate pressurisation of downstream pipework.

Where the gas supply to the meter is governed, the supply to the bypass must also be governed and, on a primary meter, the bypass valve must be sealed. On a high pressure installation the bypass valve should be gear operated to give smooth manual control of the gas flow. On rotary displacement meters, the bypass valve should be fitted at the outlet end of the bypass to avoid pulsation in the dead leg.

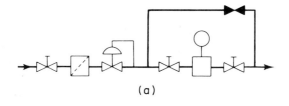

(a)

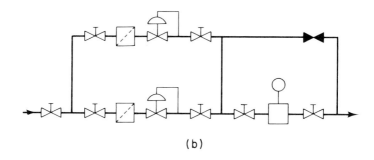

(b)

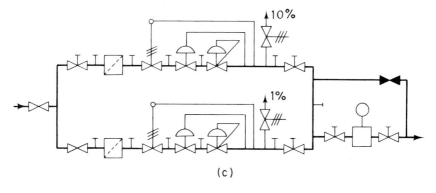

(c)

Fig. 20. Bypass and governor layouts

There are a number of possible arrangements of meters, governors and bypass. Figure 20 shows typical layouts. The first two are low pressure systems. Symbols are shown in Fig. 21.

a) Bypass to meter only.

This does not give continuity of supply when the governor is serviced or the filter exchanged.

b) Parallel governor streams.

These allow the governors or filters to be serviced whilst still metering the supply.

c) Intermediate pressure installation.

Common method is to use two governors in series with their pressure settings stepped. This is called "monitor/active" regulation. One governor (active) normally controls the outlet pressure and the other (monitor), set at a pressure 10% higher, takes over control if the first governor fails.

The slamshut valve is set 10% above the monitor.

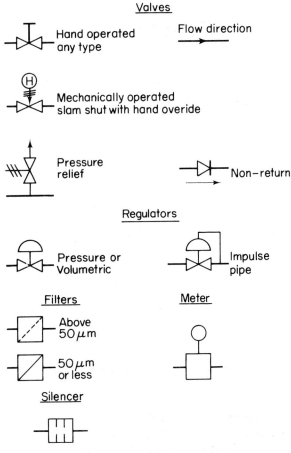

Fig. 21.   Symbols

§33   Warning Notices

Permanent notices which call attention to the special features of the installation should be mounted in a prominent position near to the meter. Their purposes are to:

- *indicate that a service syphon is fitted
- *indicate the action to be taken by the customer in the event of a gas escape
- prohibit smoking or naked lights in the vicinity of the installation
- indicate where a common service pipe supplies two or more primary meters (this must include the number of meters and preferably their location).
- *indicate the number and preferably the location of any secondary meters and give an instruction for the gas to be turned off at all appliances or meters before turning off the primary meter inlet valve
- give instructions for restoring the supply by opening the bypass valve when all appliances and secondary meters have been turned off
- prohibit the use of boosters, compressors or gas engines without permission from British Gas
- indicate that a non-return valve must be fitted in the gas supply where the gas is used in conjunction with compressed air or other gases
- warn that the meter may be damaged if the inlet valve is not fully open before starting any booster, compressor or gas engine, or if the valve is closed while the plant is running; (devices to protect the meter are described in Chapter 4).

In the case of high pressure installations, to:

- indicate the presence of gas at high pressure and any pressure limits agreed with the customer
- prohibit interference with the installation except to deal with a gas escape or open a meter bypass
- * the items indicated are statutory requirements

§34   Gas Volume Correction

The effects of changes in pressure and temperature on the volume of a given mass of gas and on gas measurement were described in Volume 1, Chapter 7.

The basic unit of gas volume is the standard cubic metre measured at 15°C, 1013.25 mbar and dry.

The basic unit of gas measurement, used in metering, is still the standard cubic foot, measured at 60°F, 30 inches mercury and dry.

Where the gas pressure is in excess of 2 bar, allowance has to be made for the compressibility factor "Z". This is dependent on the pressure, temperature and density of the gas. In orifice meter calculations a supercompressibility factor "F" may be used. This is based on American data.

Volume correction may be applied either by fixed factors or by automatic correctors. The automatic correctors may be either mechanically or electronically operated. In all cases it is usual for a badged meter to be used to provide a standard reference measurement on which calculations may be based.

Fixed factors

This method may be used with all types of meter. The requirements of the system are:

- known and constant gas metering pressure
- known average metering temperature

(a)

(b)

Fig. 22. Temperature compensated meters (a) Dresser TC (b) I.G.A. CUMTC

The pressure factor should be based on the actual pressure at maximum flow. This will be slightly lower than the pressure at low flow, due to the governor characteristics.

Temperature factor may be based either on an agreed average figure for example 16°C (50°F), or on recorded gas temperatures.

Mechanical correctors

A simple form of corrector is incorporated in the "temperature-compensated" meters. These are available as special models of some diaphragm and rotary displacement meters. An example is shown in Fig. 22.

Other separate correctors are available with either step, or continuous integrators. They may correct for temperature, or pressure or both, an allowance for compressibility factor may be built in for a given metering pressure. The system requirements are:

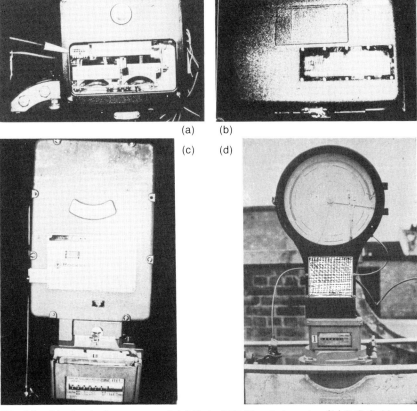

(a)   (b)

(c)   (d)

Fig. 23.   Mechanical correctors (a) I.G.A. BVI Step integrator (b) P.C.C. Mercor (c) Rockwell EM (d) Rockwell Temcorrector Type 'T'

- mechanical drive, usually from the index, giving the metered volume or electric pulses and an electro-mechanical drive
- pressure connection
- temperature sensing, usually from a "thermowell" in the gas line

The correctors may be used with rotary displacement and turbine and some high-pressure, aluminium-case meters. Examples are shown in Fig. 23.

Electronic correctors

These instruments are operated by electronic circuits. The requirements of the system are:

- electrical output from meter index
- pressure and temperature transducers

The "Z" factor for the stated working pressure may be incorporated. Examples are shown in Fig. 24.

 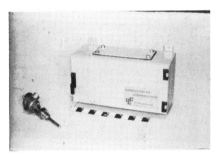

Fig. 24.   Electronic correctors (a) P.C.C. Nineteen–80 Pressure corrector
(b) P.C.C. Nineteen–81 Temperature corrector

Other systems

It is possible to use a combination of a fixed factor with an automatic correction. For example, small loads may be fully corrected economically by applying a fixed pressure factor to a temperature compensated meter reading.

Correction for pressure, temperature and compressibility can be replaced by a correction for density. The density of the gas is measured by a density cell in the gas stream. The system also requires an electrical output from the meter index and a minicomputer to give the corrected volume. Density correction may be used with rotary displacement and turbine meters. It may also be used with orifice meters, see BS 1042 : Part 1. A constant gas composition is essential, otherwise it is necessary to apply corrections for changes in S.G.

Protection from interference

Where a factor is based on an agreed pressure, the governor setting must be protected from interference. The governor and its auxiliary systems should be sealed as well as the meter bypass valve. Automatic correctors or chart recorders and also the points at which temperature and pressure readings are taken should be sealed off against interference.

§35 Isolating Valves

It should be possible to isolate any metering installation, and particularly a high pressure installation, from a safe distance.

Where valves are fitted close to meters, they should be of a type which will not cause undue flow disturbances.

Valves must have suitable characteristics for their particular purpose bearing in mind that:

- meter valves remain open for long periods and must then shut off completely on the first operation
- bypass valves remain closed for long periods and must then open quickly in an emergency

It may be advisable to use lubricated valves or valves with PTFE seatings. When lubricated valves are used upstream of the meter, care must be taken to ensure that excess lubricant is not carried into the meter where it might interfere with the operation of the meter or instruments.

It is advisable to use valves which indicate their degree of opening.

§36 Commissioning Meters

Pre-installation check

Before fitting any meter the following points should be checked:

- pipework flanges are in line, square to the pipe and the correct distance apart
- filter bodies have been cleaned out and the correct elements fitted
- upstream pipework between the meter and the filter is dry, clean and free from debris
- rotary meter rotors or impellors are free running
- meter flange faces are undamaged and meters positioned in the correct direction of flow

After fitting, check the meter and pipework supports.

Handling

All meters are precision instruments and must be treated with great care, they must never be dropped or mishandled. Always lift in accordance with manufacturer's instructions, particularly when using slings and mechanical lifting aids.

Avoid strain on the index and damage to the index glass. Replacing a glass may require a visit from the official meter examiner.

During transport or storage the meter connections should be sealed off.

Testing

For installations operating at pressures below 75 mbar (30 in w.g.) the meter installation should be tested for soundness in accordance with the current edition of "Soundness Testing Procedures for Industrial and Commercial Gas Installations", published by British Gas.

Purging

The procedure given in B.G.C. "Purging Procedures for Non-Domestic Gas Installations" should be followed.

Provision for purging should be provided in the pipework. Vent points should be fitted between isolating valves so that meters, filters and governors may be purged individually.

Small installations may be purged using the gas itself. For installations with pipes of above 100 mm (4 in) bore, an inert gas such as nitrogen should be used.

Commissioning

All meters should be pressurised slowly. The procedures for commissioning high pressure meters are given in IGE/GM/1, 1974 and vary with the type of installation. Manufacturers' instructions should be carefully followed.

Where possible, the bypass should be used to pressurise downstream pipework. The meter should be pressurised through the outlet valve, in the case of diaphragm meters and the inlet valve on rotary and orifice plate meters. Follow this operation by opening the other meter valve and closing and sealing the bypass valve.

The index should be checked to ensure that the meter is operating. Rotary displacement meters should also be checked for noisy operation or high differential pressures which indicate that the meter is not operating freely. The oil sump levels should also be checked.

Exchanging

Before commencing any work ensure that the replacement meter is correct with respect to:

- badged rating
- operating pressure
- case dimensions
- matching flanges or other connections

Ensure maximum ventilation and no naked lights in the vicinity. Check that breathing apparatus and other safety equipment is available, as required.

With rotary displacement meters check the new meter for clean rotors, and freedom of internal parts. Check free rotation of turbine meters. To exchange:

- open the bypass valve or fit temporary rider
- close the meter isolating valves and check that the outlet pressure is maintained
- check that inlet and outlet valves are holding
- on large or high pressure installations, vent and purge the meter; small meters may be purged by venting in a safe area after removal
- remove the meter and seal the inlet and outlet connections
- install the replacement meter
- purge the meter with the inert gas if necessary
- check the meter and connections for soundness
- open the inlet valve and pressurise the meter slowly
- open the outlet vent point and purge the meter of air or inert gas
- close the vent and open the meter outlet valve
- close the bypass valve and check that outlet pressure is maintained

- check the meter for noise and smooth operation of index
- seal the bypass valve

A temporary continuity bond should always be attached before a component or any pipework is disconnected. The bond should remain in position until all the connections have been remade.

Industrial Processes and Plant

Chapter 3 is based on an original draft by Mr. A. J. Spackman

PROCESSES

Gas is used in many industries for a wide variety of processes whose temperatures range from about 70°C (160°F) up to 1650°C (3000°F). At the lower end of the scale are the drying operations carried out on paints, inks and a number of granular materials including foods and pharmaceuticals. At the upper end, high temperatures are required for melting metals and glass and for heating metals prior to hot working.

§37 Drying

Drying is the removal of water, or some other solvent, from a product. It may be carried out by heating the product to about 70 to 90°C (160 to 190°F) and allowing the solvent to evaporate freely. This takes a long time, so commonly radiant heating or forced convection are applied. In forced convection, hot air and sometimes products of combustion from an air heater are circulated through and over the product. Water vapour or solvent is taken up in the air, so drying the product. The rate of drying depends on the:

- temperature of the hot air
- input and output humidities
- rate of hot air circulation

In some cases moisture is removed and the air recycled, as in an air conditioning plant. Where toxic or flammable solvent vapours are involved special precautions, specified by the Factory Inspectorate, must be taken.

Some materials, including paints, enamels, oils and resins may undergo chemical changes when heated to temperatures of about 120 to 240°C (250 to 460°F). The processes are known as "curing", "stoving" or "baking" and they produce harder surface coatings than are obtained with normal air dried paints.

Cores made of sand for foundry work are dried at 120 to 240°C (250 to 460°F) and the moulds require temperatures of up to 200 to 400°C (390 to 750°F).

§38 Cooking

This is usually considered to be a "commercial" operation rather than an "industrial" one. However, the bulk baking of bread, biscuits, pies and pastry may be carried out in large factories and often in travelling ovens 30 to 60 m long. Typical temperatures are:

- cakes 150 to 180°C (300 to 350°F)
- biscuits 190 to 240°C (375 to 460°F)
- bread 150 to 260°C (300 to 500°F)
- pies about 260°C (500°F)

The confectionery trade produces sweets at temperatures of 110 to 150°C (230 to 300°F) while potato crisps and chips are fried at about 180 to 210°C (350 to 410°F).

Although not a food, varnish is "cooked" or boiled in a special pan with a fume offtake at temperatures of 270 to 320°C (520 to 610°F).

§39 Vitreous Enamelling

This process was described briefly in Volume 1, Chapter 11. It consists of first preparing the surface of the metal by degreasing and by dipping in an acid solution to etch it so that the frit is keyed on.

The frit, which is composed of the glazing materials mixed with water and some clay, is applied to the metal by dipping or spraying. The components are then dried in an oven.

When dried, the frit is fired in a furnace at 650 to 700°C (1200 to 1290°F) for cast iron components and 760 to 930°C (1400 to 1710°F) for steel plates.

§40 Ceramics

The pottery industry has turned from the use of coal to the refined fuels, including gas. Most pottery is fired twice. The first or "biscuit" firing is to vitrify the clay and the second or "glost' firing is to fire on the glaze. Glazed sanitary ware may be produced by a single firing.

The products are first dried to drive off a large proportion of the moisture. This may be in a separate oven or in the preheating section of a continuous kiln.

The temperatures for firing are:

- biscuit—earthenware 1150 to 1250°C (2100 to 2280°F)
  —porcelain 1250 to 1400°C (2280 to 2550°F)
- glaze 1050 to 1150°C (1920 to 2100°F)

Muffle furnaces were commonly used for pottery to protect the ware, but with natural gas, open furnaces may be employed.

Bricks are fired once at about 990 to 1430°C (1810 to 2610°F).

§41   Glass

There are many different kinds of glass, each made for a particular purpose. The main constituent is silica, or sand, but various other ingredients are added to obtain the different properties. The three most common forms of glass are:

- crown glass—silica 72%, soda 15%, lime 9%, oxides 4%
- flint glass—silica 55%, lead oxide 32%, potassium and sodium oxides 12%, other oxides 1%
- heat resisting glass—silica 74%, boric oxide 16%, potassium sodium and aluminium oxides 10%

Crown glass is an easily worked glass with a low melting point and is used for windows, bottles, electric lamp bulbs and tubes.

Flint glass, also called "lead crystal" is a fine, clear glass used for optical lenses, table glasses and decorative ware. It is also used for cut glass work and engraving.

Heat resisting glass, or boro-silicate glass has a high melting point and a low coefficient of expansion. It may be made thicker than ordinary glassware to give strength to ovenware dishes.

§42   Melting

Glass is melted in tanks or in pot furnaces at temperatures of 1300 to 1600°C (2370 to 2910°F). Tank furnaces may contain up to several hundred tonnes of soda glass, supplying a float glass or a bottlemaking plant. Small kilns may heat one or more pots containing about 50 kg of glass for scientific or lead crystal glassware. The basic operations in manufacturing glass are:

- measuring and preparing the ingredients, these include some broken glass or "cullet"
- melting the batch, known as "founding"
- heating to above melting point to remove the bubbles of trapped gases and refine the glass, known as "fining"
- conditioning the glass by reducing the temperature to that required for working

In a tank furnace, the operation is continuous with the frit and cullet fed in at one end and the conditioned glass withdrawn at the other. The burners are arranged to give the temperatures required at each stage of manufacture. Pot furnaces are intermittent.

§43 Fabricating

The methods used for producing glass articles are:

- blowing
- pressing
- rolling, drawing or floating
- cooling in the pot

Blowing

Glass blowing was invented by the Romans and hand blowing is still used for high quality articles in small numbers. The required quantity of glass is withdrawn from the pot on the end of a 1.5 m (5 ft) tube. By controlled blowing, rotating and swinging in an arc, the glass-blower produced the desired shape. This may then be placed in a hinged mould and blown to fit the mould. Other processes may be necessary to modify the shape and add or remove parts. Glass stems or handles may be attached and unwanted glass cracked off by scratch marking and applying heat. Gas flames, sometimes with oxygen, are used for these operations. After cracking off, the surface may be ground or fire-polished to give a smooth, even finish.

Bottles and electric light bulbs are blown on automatic machines which simulate the actions of the glass blower. The glass is fed from the tank down to the machine through a channel or "feeder" which is heated to ensure that the glass reaches the machine at the correct working temperature. Air blast burners are frequently used for this purpose. The "gob" of glass from the furnace is first hollowed by a plunger in the "parison mould" and then blown to shape in the final mould. The machines are equipped with groups of gas burners to reheat, cut and polish the products as required.

Pressing

Pressing may be carried out by hand but is now commonly used for the manufacture of domestic hollow-ware, car headlamp glasses and similar products by automatic machines. The gob of glass from the furnace is dropped into a mould and squeezed into shape by a plunger. The moulds are positioned around the circumference of a rotating table so that each mould, in turn, receives its glass, is pressed and is then finished and emptied.

Rolling, drawing and floating

These are all methods of producing sheet glass, the bulk of which is used for window glazing. Figured glass is produced from patterned rollers and, in reinforced glass, wire netting is introduced before rolling. Special glass is now made to filter sunlight so that shades and blinds are unnecessary.

Cooling in the pot

This method is used for the manufacture of optical glasses. The contents of the pot are stirred after fining and removed to another furnace for slow cooling. When cold, the melt is broken up and the pieces examined for flaws. The selected glass is then reheated. moulded and cooled slowly again. It is then again examined for defects and only perfect pieces are given the final processing.

§44   Annealing

When the fabrication is finished, glass must be annealed to remove the stresses set up by uneven cooling. This is carried out in an oven, called a "lehr" at temperatures of about 450 to 650°C (850 to 1200°F). The lehr may be a simple box oven or a long, travelling oven, depending on the quantity and the time required for treatment. Some optical lenses may need several months to cool down to prevent distortion and stress.

Glass which has not been properly blown or annealed may have built in stresses which will cause it to shatter, often for no apparent reason. This property is used in the manufacture of "toughened glass" for motor car windscreens and windows. The outer surface of the glass is heated to red heat and then cooled rapidly by air jets or by dipping in oil. This sets up high comparative stresses, and, when broken, the glass fractures into small chunks, instead of splinters.

§45 Heat Treatment of Metals

A simple explanation of heat treatment is given in Volume 1, Chapter 11.

Metals are given heat treatment in order to bring about changes in the structure of the metal. These changes may be limited to the surface structure or may go right through to the core of the object. The changes are brought about by heating and cooling to a specified programme appropriate to the properties of the metal and the purpose for which the article was designed.

Heat treatment is carried out in a variety of furnaces and baths at temperatures from about 200°C to 1000°C (390°F to 1830°F). The furnaces may be either directly or indirectly fired and the baths may contain oil, molten salts or lead. Fluidised beds may also be used for heating or cooling the products.

Working flames are used for treating some articles made from iron or steel. By applying a flame directly to the surface, or a specific area of the product, that part may be hardened, to resist wear, whilst the core remains tough, to withstand loads and shocks. For example, a gear wheel must be tough, but its teeth must be hard.

Gear wheels are flame hardened by mounting them on a rotating table over a quenching bath of oil. A number of burners are positioned so that, as the wheel rotates, the faces of each tooth are evenly heated. When the required temperature is reached, down to an appropriate depth, the gas is reduced to a low rate and the wheel and table automatically lowered into the quenching bath. The main core of the wheel does not get hot enough to change its characteristics.

There are four main methods of flame hardening:

- spot method, in which a specific area is heated and then quenched, used for rocker arms and chain links
- spinning method, where the component is rotated between a number of burners and then lowered into the quench bath, used for gear wheels
- progressive method, in which the flame head moves over the article, followed by the quenching medium; used on slideways of machine tools, wheel tyres and rims and large gear teeth
- a combination of rotation and progression used on spindles and rolls

§46 Ferrous Metals

The iron-carbon diagram, Fig. 1, shows the temperatures at which changes occur in the structure of the various forms of iron or steel.

These temperatures are known as the "critical temperatures" and the diagram represents the relationship between the composition of the metal and its critical temperatures under very slow heating and cooling conditions. In addition to the phases shown on the diagram, steel can also exist in other forms, one of which is "martensite". This

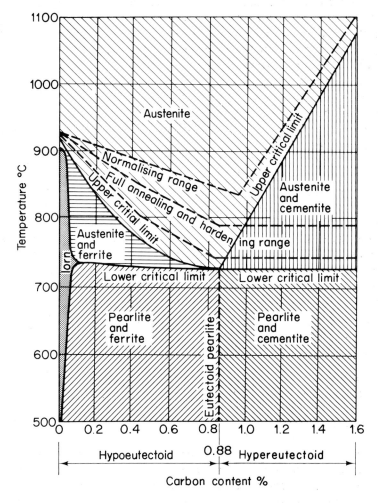

Fig. 1. Iron-carbon diagram

is very hard and brittle and is formed when austentite is cooled quickly from above the critical temperatures. The process may be reversed by heating the martensite and cooling it slowly. So the formation of martensite makes an important contribution to the hardening and tempering of steel.

The principal heat treatment processes are:

* annealing
* normalising
* hardening
* tempering
* case hardening

Annealing

Annealing is a softening process which is carried out at temperatures of between 700 and 900°C (1290 and 1650°F). It consists of heating to the required temperature followed by slow cooling, usually in the furnace. Annealing may be either "full annealing" which entails heating the metal to above its lower critical temperature, or "sub-critical annealing" when it is heated to just below the critical temperature.

Full annealing improves some of the qualities of steel by refining the grain structure.

Subcritical annealing is used for stress relieving lower carbon steels which have been welded or cold worked.

"Spheroidising" is a form of prolonged annealing at, or slightly below, the critical temperature, which improves machinability.

Normalising

This process is similar to annealing but is carried out at higher temperatures, 900 to 1100°C (1650 to 2010°F). The steel is heated to above its upper critical temperature and then cooled in air. The slightly faster cooling rate is necessary to maintain a satisfactory grain structure.

"Patenting" is a method of softening steel wire which has been work hardened by drawing. It entails heating to about 1000°C (1830°F) and quenching in air or in a lead bath at 500 to 600°C (930 to 1110°F).

Hardening

Hardening consists of heating the steel to above its critical temperature, followed by quenching quickly. Water is used for fast quenching. Generally, faster quenching produces harder steel. Hardening produces martensite in the steel which renders it hard and brittle. For most practical purposes the steel then needs to be tempered.

Typical temperatures for hardening are 750 to 850°C (1380 to 1560°F).

Tempering

Tempering consists of heating the hardened steel to temperatures of 150 to 640°C (300 to 1180°F), below the critical temperature. This converts the martensite into less hard but tougher substances. The rate of cooling varies with the results desired and the composition of the steel. Tempering may be carried out in furnaces, oil, lead and salt baths.

Austempering, martempering and isothermal quenching are processes of interrupted quenching used to develop toughness and ductility and prevent internal stresses in the steel.

Case Hardening

Low carbon steels which cannot be hardened by normal heat treatment may be given a "case" of surface skin which has been suitably altered in composition to respond to hardening. This is done usually by:

- carburising
- nitriding
- carbo-nitriting
- cyaniding

Carburising is a process in which carbon is introduced into the surface of the metal to a depth of 0.25 to 1.5 mm (0.001 to 0.06 ins). This is by heating the metal to above the critical temperature while it it is in contact with a material containing carbon. The material may be solid, liquid or gas.

The old method was to pack the work in solid charcoal in steel boxes and then heat it to about 900°C (1650°F) in a furnace for several hours.

A quicker method is to use molten salts containing sodium cyanide, known as "cyaniding". This gives better uniformity of heating and needs shorter periods of immersion.

Gas carburising is commonly used with hydrocarbons in the form of propane or natural gas being introduced into the furnace atmosphere. Temperatures used are in the region of 950 to 1050°C (1740 to 1920°F).

Nitriding is the formation of a hardened case by the addition of nitrogen to the surface, without quenching. Carbo-nitriding uses a controlled gas atmosphere with the addition of ammonia gas so that both carbon and nitrogen are absorbed. Temperatures are typically 850 to 950°C (1560 to 1740°F).

After carburising, the work must be heat treated to harden the outer case and to refine the core. The whole process of forming and hardening the case is termed case hardening.

§47   Non-Ferrous Metals

The main heat treatments of non-ferrous metals are annealing and ageing. Typical temperatures are given below.

Annealing

Copper is annealed at 430 to 650°C (810 to 1200°F); the major problem in the process is the prevention of oxidation and copper wire may be bright-annealed in a vacuum. Another simple method is to pass the wire in a continuous strand through a short refractory tunnel fired by a single gas burner. No excess air is permitted to enter and the wire is cooled by a stream of water as it leaves the tunnel.

     Brass is annealed   at 320 to 480°C (610 to 900°F)
     Nickel and Monel   at 600 to 800°C (1110 to 1470°F)
     Aluminium   at 220 to 400°C (430 to 750°F)

Ageing

Both aluminium and magnesium alloys may be hardened by ageing. This is carried out at low temperatures. Generally increased hardness is obtained by immersion for longer periods at lower temperatures. Typical temperatures are:

- aluminium   120 to 230°C (250 to 450°F)
- magnesium   180 to 200°C (360 to 390°F)

§48   Hot Working Metals

Hot working of metals includes:

- forging
- drawing
- extruding
- rolling

For most of these processes the metal is heated in a furnace and it may be necessary to reheat the work between operations. Many different types of continuous and intermittent furnaces are used depending on the size, shape and weight of the work.

Temperatures required for hot working are typically:

- aluminium   430 to   500°C (810 to 930°F)
- brass   680 to   780°C (1260 to 1440°F)
- steel   1100 to 1300°C (2010 to 2370°F)

§49 Melting Metals

The metals most commonly melted by gas are:

- tin   260 to   340°C (500 to 640°F)
- typemetal   270 to   340°C (520 to 640°F)
- lead   330 to   390°C (630 to 730°F)
- zinc   430 to   480°C (810 to 900°F)
- aluminium   650 to   760°C (1200 to 1400°F)
- magnesium   680 to   700°C (1260 to 1290°F)
- brass   930 to   980°C (1710 to 1800°F)
- copper   1150 to 1250°C (2100 to 2280°F)

All these metals oxidise readily at their melting temperatures and this may need to be avoided to reduce the loss of metal and prevent any included oxide spoiling the work.

In addition the molten metals absorbs some of the gases in the furnace atmosphere. This can result in blow-holes in the casting or changes in the structure of the metal.

Some metals have particular hazards, for example, lead vapour is toxic and magnesium and aluminium can be burnt.

For melting metals at temperatues below 500°C (930°F) the pot furnace is generally used and immersion tubes are also employed.

At temperatures above 500°C, the furnaces used are the crucible and the reverberatory types.

§50 Flame Processing

In addition to their use for flame hardening and glass working, gas flames may also be employed for:

- flame cutting and profiling
- soldering and brazing
- metal spraying

Flame cutting

This is similar to oxy-acetylene cutting but with gas replacing the acetylene. Compressed air may be used instead of oxygen for some applications. On profiling machines, a sensing head follows a pattern or a computer programme and directs the cutting head on the work.

Soldering and Brazing

Soldering may be carried out by manual or automatic soldering irons, by dipping in molten solder or with a variety of blow pipes and torches. Tin cans and boxes are produced on continuous conveyors by automatic methods. The gas burners may be low or high pressure and may use compressed air.

Brazing is similar to soldering except that it requires a higher temperature, about 700°C (1290°F). Torches usually employ high pressure air. Items such as heat exchangers and car radiators are brazed in a travelling oven, the brazing metal being included at the assembly of the component.

Metal spraying

Many non-ferrous metals may be spray coated on to other metals, including tin, lead, zinc, aluminium and copper. The process involves passing the metal, in a finely divided form, through an oxy-gas flame at high velocity. The heated particles are deposited on the surface to be treated by a separate stream of high velocity air. The process may be used to build up worn areas of high value articles like steam valves.

§51 Submerged Combustion

This is a variation of the working flame and is used for tank heating. The flame burns below the surface of the liquid and the heat is transferred by the bubbles of the products of combustion in direct contact with the liquid.

Under atmospheric pressure the maximum temperature reached is about 82°C (180°C). The large surface area of the bubbles gives a high rate of evaporation which helps to produce a more concentrated liquid.

This method is particularly suitable for heating corrosive liquids and those liable to deposit suspended solids or scale.

PLANT

§52 Ovens and Driers

Box Ovens

A box oven is the simplest form of industrial oven. It consists of an insulated, sheet metal box with a door, heated by a burner in the base.

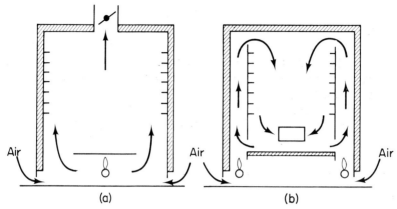

Fig. 2. Double cased oven (a) top flue (b) bottom flue

It may be either:

- double cased, that is, directly fired; or
- treble cased, that is, indirectly fired.

In the double cased oven, Fig. 2, the products of combustion circulate within the oven and are removed by the flue, together with moisture or any volatile materials from the product. The flue is often taken from the top of the oven but it can also be situated at low level in the rear wall to give improved circulation and a more even oven temperature, Fig. 2(b).

The rate of ventilation is usually controlled by a flue damper. This is designed so that, when it is closed, at least one third of the flue area still remains unobstructed.

In some industrial processes, it is not possible to use the products of combustion as the means of transferring heat directly to the work. For these processes the treble cased oven is used, Fig. 3. This has an inner box with its own separate air inlet, flue outlet and appropriate ventilation control. The walls of the box are heated by the products circulating around them and so form the heat transfer surfaces. In a number of these ovens, fans and baffles are used to circulate the air over the walls of the oven and so increase the rate of heat transfer, Fig. 3(b).

Box ovens were generally heated by bar burners and fitted with thermo-electric flame monitoring systems. Electronic systems are now becoming more common, due to their more rapid response times.

The flue system is designed to give maximum efficiency. The flue damper may be automatically controlled by the gas control system. A motorised damper can be opened before the gas rate is increased and closed after the gas rate is reduced. This adjusts the flow through the flue to that required for the particular quantity of gas being burnt at that time, so limiting the volume of excess air entrained.

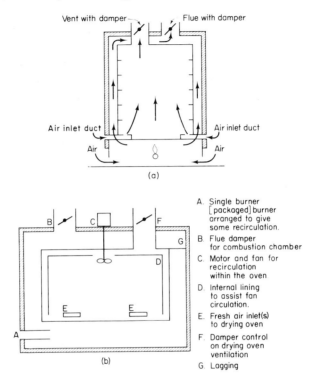

Fig. 3.   Treble cased oven (a) simple box oven (b) fan assisted air recirculation

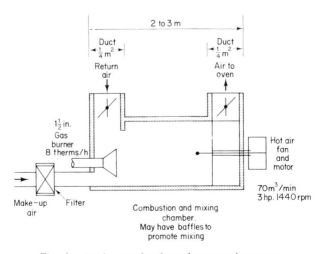

Fig. 4.   Air heater for forced convection oven

Forced Convection Ovens

Independent, direct fired air heaters are now commonly used to provide forced convection for drying and stoving. They can be sited in an otherwise unused space and connected to the oven by ducting. Alternatively, they may be mounted above or below the oven itself. Existing ovens may be converted to forced convection. Efficiency can be increased by recirculating a proportion of the products of combustion. The general arrangement of an air heater is shown in Fig. 4.

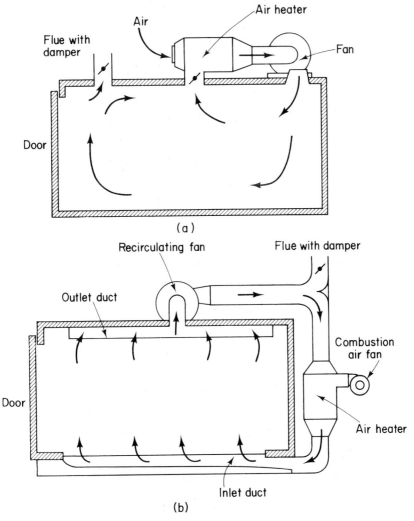

Fig. 5. Forced convection ovens (a) suction burner (b) pressure burner

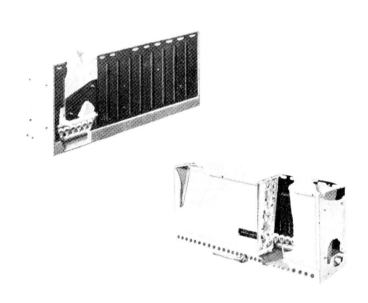

Fig. 6. Low temperature infra-red panel

Fig. 7.   Arrangement of panels in drying oven

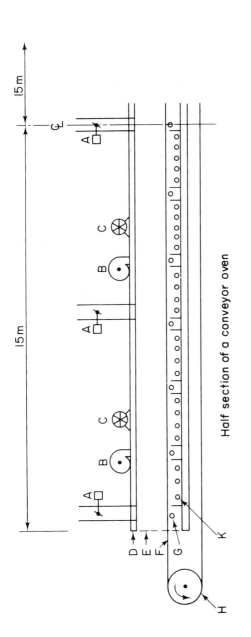

Half section of a conveyor oven

A. Motorised damper in flues.

B. Air fan.

C. Gas compressor.

D. Insulated case.

E. Optional curtain.

F. Conveyor, steelband, or mesh.

G. Idler rollers for conveyor.

H. Drive roller for conveyor.

K. H.P. Gas burners. All air supplied from duct within D.

Burners fitted with individual ON/OFF taps.

Burners grouped for gas and air supplies and flue.

Thermostat control operates flue damper to suit gas flow.

Fig. 8. Horizontal conveyor oven

Recirculation ovens have integral air heaters which may be either:

- pressure type, with the fan on the inlet, or
- suction type, with the fan on the outlet.

Examples are shown in Figure 5.

Radiant Panels

Drying processes may be carried out using infra-red radiation from gas fired radiant panels. These can be obtained with emissivities to match the particular materials to be dried.

Panels may be either:

- high temperature, giving temperatures up to 1000°C (1800°F).
- low temperature, giving temperatures up to 350°C (660°F).

High temperature panels use the radiant burners with porous or perforated ceramic tiles described in Volume 1, Chapter 3.

Low temperature panels usually have a black cast iron, or vitreous enamelled steel panel heated from behind by a bar burner, Fig. 6.

Because radiation travels in straight lines, these panels are best suited to drying coatings on flat products. Banks of panels may be built up into tunnels through which the products travel on a conveyor. Typical arrangements are shown in Fig. 7.

Conveyor Ovens

These are generally either:

- horizontal, or
- camel-back

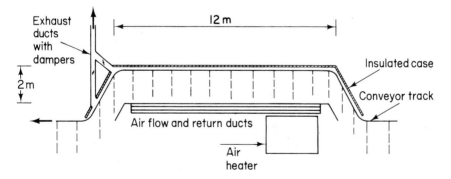

Air heater placed in space below oven.
Flow and return air ducts placed to suit
required flow and temperature distribution.

Fig. 9.   Camel-back oven

The horizontal oven, Fig. 8, has a flat metal conveyor, often of wire mesh, passing above gas burners. The oven illustrated is a type used for baking bread, pastries or biscuits. It is fired by high pressure gas burners which entrain combustion air from a duct which surrounds the oven. These ovens, which may be up to 60 m (200 ft) in length, are divided into zones, with its own compressors and flue. The zones may be operated at different temperatures by individual thermostats. The camel-back oven, Fig. 9, is widely used for drying and stoving in light industries. The entry and exit are sloped downwards at 45° forming seals to retain the hot air. The products are carried through the oven suspended from an overhead conveyor. The air heater which provides forced convection with recirculation is located below the centre section.

Dryers for Granular Materials

A number of granular materials such as sand, chemicals and food-stuffs, are processed in various dryers using hot air from a direct fired air heater. The common forms of drier include:

- · rotary driers
- · spray driers
- · flash driers

Rotary driers are similar to tumble driers. The rotating shell may be inclined so that the product moves from the feed hopper to the other end by gravity. As the shell rotates, the inner paddles or "flights" continually turn the particles over in the stream of warm air which often travels counter to the feed.

Spray driers have the product fed in as a liquid or a slurry. It is then atomised or sprayed and dried in a stream of hot air.

Flash drying consists of dropping the wet particles into the stream of hot air and removing them when dry, usually by a cyclone through which the air passes on its way to the flue.

§53 Tanks and Baths

The terms, "tanks" and "baths" are used, almost interchangeably, to describe a variety of containers, operating over a wide range of temperatures. What is called a "tank" in one industry is called a "bath" in another. They vary from the rectangular, galvanised iron tank holding a solution of caustic soda in water, to the small steel pot in a refractory setting for melting printer's type metal on a lino-type machine.

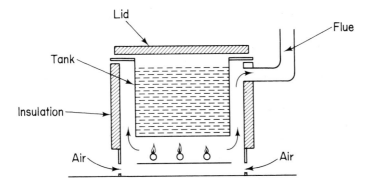

Fig. 10. *Underfired tank*

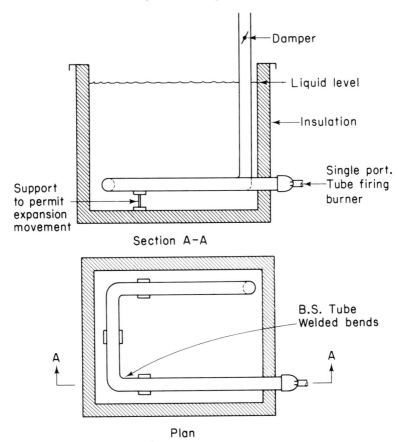

Fig. 11. *Immersion tube heated tank, for liquids and metal melting. (Length and diameter of tube required and hence configuration depends on the desired heat input and space available for the tubes.)*

Underfired Tanks

These are heated by burners below the tank, Fig. 10. The products of combustion pass around the sides and discharge at the top, usually into a flue. The top of the tank should have a cover to retain the heat. Low temperature tanks may be insulated by floating plastic spheres on the liquid surface.

Salt and metal baths must be evenly heated to avoid the problem of solid material existing on top of molten material which is expanding and which could cause an explosion.

Immersion Tubes

The tubes may be from 50 to 150 mm (2 to 6 in) diameter. They may be heated by natural draught burners but increased efficiencies and higher gas rates can be obtained with forced draught systems.

The length and layout of the tubes depends on the size of the tank and the heat input required. A typical layout of a single tube is shown in Fig. 11. The "Temgas" immersion heater may be fitted to existing tanks without cutting into the sides, Fig. 12. It has a combustion chamber with internal baffles, in place of the usual tube or tubes.

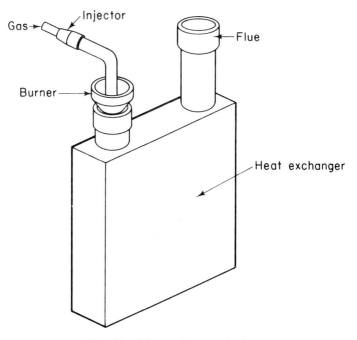

Fig. 12.   'Temgas' immersion heater

Submerged Combustion

A form of submerged combustion burner is shown in Fig. 13. Because the combustion must be completed and the products expelled against the pressure of the liquid, this burner uses a premix machine to obtain its air/gas mixture. Some nozzle mixing burners are used but premix systems are more common.

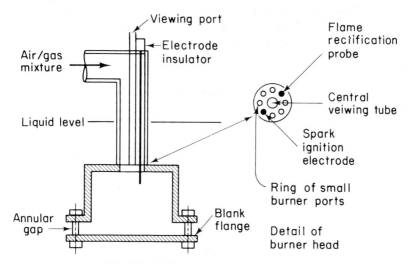

Fig. 13.   Submerged combustion burner

FURNACES

54   Direct Fired

Reverberatory Furnaces

A reverberatory furnace is strictly one in which the heat is reflected down onto the surface of the molten metal or glass from the crown of the furnace. The name is however applied to any open hearth furnace in which the burners fire onto the metal, Fig. 14.

These furnaces are commonly used for remelting large amounts of scrap metal and may hold up to 50 tonnes. Nozzle mixing burners are generally used.

Oven and Box Furnaces

These are simply refractory boxes in which the work is placed on the hearth, which forms the floor of the furnace, Fig. 15. "Oven" fur-

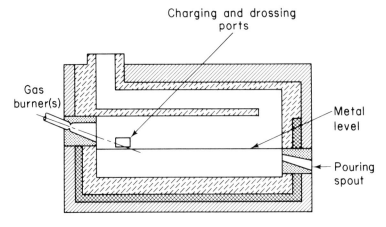

Fig. 14.   Reverberatory furnace

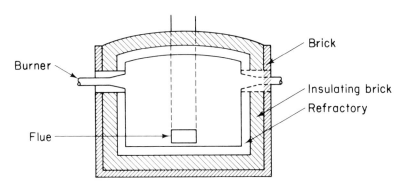

Fig. 15.   Box furnace

naces are generally smaller than box furnaces and have the hearth at waist height for easy loading and unloading. Access to the furnace is usually by a counterbalanced, lift-up door. The furnaces are commonly fired by nozzle mixing burners. There are many variations of these furnaces used as kilns and for heat treatments of metals and glass.

Bogie Hearth Furnaces

In order to make loading and unloading easier and to save time waiting for the furnace to cool down so that the load may be handled, these furnaces have a hearth which may be withdrawn on rails, Fig. 16. If two hearths are used, the load on one may be removed and replaced while the second is being processed.

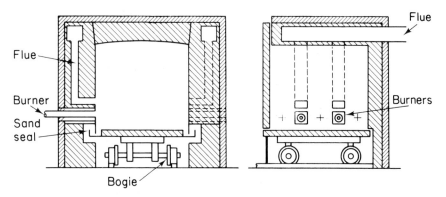

Fig. 16.   Bogie hearth furnace

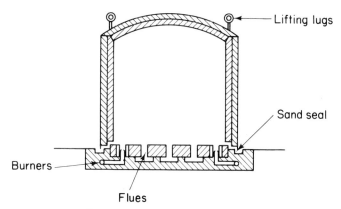

Fig. 17.   Portable cover furnace

Portable Cover Furnaces

These use another method of making loading and unloading easier. The whole top of the furnace may be lifted off the hearth by an overhead crane. The base of the furnace usually contains the burners and the flue ducts, Fig. 17. Two hearths may be used with one cover which is fitted over the hearths alternatively, as they are reloaded.

§55   Indirect Fired

Crucible Furnaces

Crucibles are metal melting furnaces used where direct firing methods would affect the metal or when only small quantities of metal are to be melted. The crucible may be of either refractory material or metal and is emptied by:

- lifting out and pouring manually
- baling out by hand ladle or internal pump
- mechanically tilting

A typical lift-out furnace is shown in Fig. 18. These furnaces are commonly fired by air blast tunnel burners firing at a tangent into the annulus between the crucible and the lining. Combustion products are vented through the lid. Lift-out furnaces may be sunk into pits with their lids at floor level.

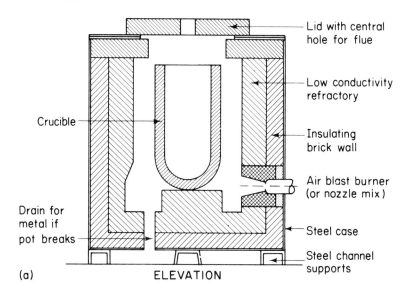

(a) ELEVATION

Lid with central hole for flue

Low conductivity refractory

Crucible

Insulating brick wall

Air blast burner (or nozzle mix)

Drain for metal if pot breaks

Steel case

Steel channel supports

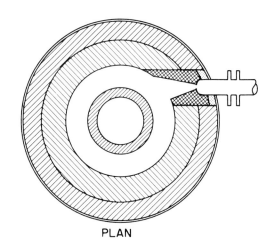

(b) PLAN

Fig. 18. Crucible furnace

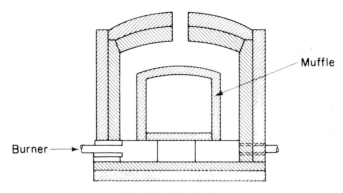

Fig. 19.   Muffle furnace

Muffle Furnaces

In these furnaces the work is protected from the effects of the products of combustion by being contained in an inner case or "muffle". In high temperature furnaces the muffle is of refractory material but below 730°C (1350°F) either cast iron or alloy steel may be used. Fig. 19 shows a small muffle furnace, fired by air blast burners. Natural draught burners may also be used. The muffles may be fed with separately produced atmospheres to prevent oxidation or to change the surface structure of the work.

A semi-muffle furnace is of similar construction but has no crown on the muffle.

Radiant Tube Furnaces

A radiant tube is an internally fired tubular heat exchanger similar to the immersion tube used in tank heating, but operating at much higher temperatures in the incandescent range. It passes through the furnace chamber and may produce furnace temperatures of 300°C

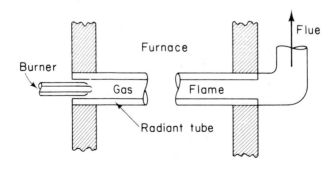

Fig. 20.   Radiant tube

(570°F) to 1050°C (1920°F). The radiant tubes may be of alloy steel or refractory materials and are located to achieve the temperature distribution required within the furnace.

A typical radiant tube furnace is shown in Fig. 20. Recent developments have produced single-ended tubes employing internal recirculation of products and recuperators.

§56   Regeneration and Recuperation

Some of the heat in the combustion products leaving a furnace may be recovered by using it to heat water or produce steam in waste-heat boilers or to dry or preheat the work. It may, however, be returned directly to the furnace by using either:

- regenerators, or
- recuperators

Regenerators

A regenerator is a series of passages through a mass of refractory material, usually chequer brickwork. Hot combustion products are passed through the regenerator, heating up the brickwork. When an adequate temperature is reached, the products are switched to a second regenerator and the incoming combustion air is heated by passing it through the hot brickwork. Fig. 21 shows an example of the layout of a regenerative furnace. There are duplicate fuel burners

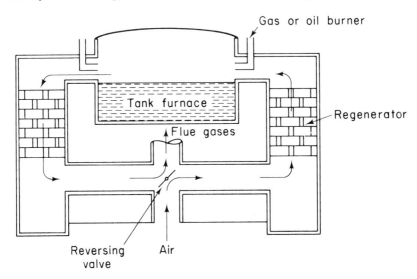

Fig. 21.   Regenerator

on both sides of the tank and these are lit alternatively as the air and flue gas circulation is changed over. Regenerators are cyclic in operation although some continuous regenerators have been produced. These employ rotating drums or moving columns of ceramic pebbles.

Recuperators

A recuperator is a form of gas-to-air heat exchanger which transfers heat continuously from the combustion products to the incoming combustion air. It may comprise systems of concentric alloy steel tubes using parallel or counterflow streams of gases and air, Fig. 22.

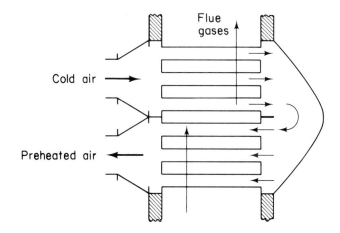

Fig. 22.   *Cross flow recuperator*

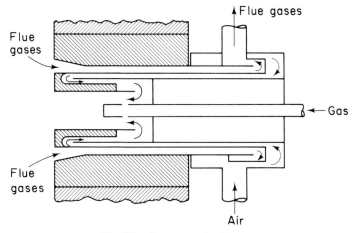

Fig. 23.   *Recuperative burner*

For temperatures above 1000°C (1830°F) refractory sections are required.

Both regenerators and recuperators are confined to use on the larger furnaces. To recover heat on smaller and often less efficient furnaces, the recuperative burner was designed. This is a nozzle mixing burner fitted with a counterflow heat exchanger, Fig. 23. The combustion products are drawn in from the furnace and pass down an annulus and over the surface of an annular tube carrying air to the burner nozzle. The products are normally extracted by means of an eductor on the boiler flue which also allows the furnace pressure to be controlled.

§57 Furnace Atmospheres

Oxidising, neutral and reducing atmospheres and their effects on iron were briefly described in Volume 1, Chapter 2. However, the terms are misleading since, as was pointed out, atmospheres which do not contain O_2 may still be oxidising if they contain CO_2 or H_2O.

In consequence, furnace atmospheres are usually described as "endothermic" or "exothermic" which relate to the methods by which they are produced.

Chemical reactions which take in heat are endothermic.

Reactions which give out heat are exothermic.

Endothermic atmospheres are produced by heating the fuel gas with a small amount of air in a retort containing a catalyst. This is known as "catalytic reforming". It produces a gas which contains principally CO, H_2 and N_2. With the addition of some propane or methane it may be used for gas carburising.

Exothermic atmospheres are produced by the combustion of the fuel gas with a controlled amount of air usually in a premix system. "Lean" atmospheres are produced by burning the gas with its theoretical or "stoichiometric" air requirement, or very slightly below. This produces a gas consisting principally of CO_2, H_2O and N_2. "Rich" atmospheres are produced by partial combustion in lower air-gas ratio mixtures.

Both lean and rich exothermic atmospheres may be "stripped", that is have the CO_2 and H_2O removed, by several different processes.

The composition and application of endothermic and exothermic atmospheres is given in Table 1. As can be seen, rich stripped exothermic gas can be used as a substitute for endothermic gas for many purposes. This is an advantage, since endothermic gas is more difficult and more expensive to produce.

§58 **TABLE 1 Controlled Furnace Atmospheres Produced from Natural Gas**

	Atmosphere	Production	Composition % Volume						Typical Applications
			CO	H_2	CO_2	H_2O	CH_4	N_2	
Exothermic	Lean	Complete combustion of fuel gases	0–3	0–4	12–10	2–3	Nil	Balance	Bright and clean annealing copper, nickel, brasses, aluminium
	Lean Stripped	CO_2 and H_2O removed	0–3	0–4	Nil	Nil	Nil	Balance	Annealing. Carrier gas for carbon restoration. Ferrous metals
	Rich	Partial combustion of fuel gas	9–12	11–15	7–5	2–3	1–2	Balance	Normalizing ferrous metals brazing and sintering copper
	Rich stripped	CO_2 and H_2O removed	10–13	12–15	Nil	Nil	1–2	Balance	Substitute for endothermic atmosphere for most purposes
	Modified	Partial combustion $CO_2 + H_2O$ removed. CO shift reaction CO_2 and H_2O removed	Nil	3–12	Nil	Nil	Nil	Balance	Long cycle annealing, low carbon and mild steels
Endothermic	Endothermic	Catalytic reforming fuel gas and air	20–25	30–45	Nil	Dewpoint +15 to −15	0.5 to 1.0	Balance	Hardening, brazing + sintering carrier gas for carburising and carbonitriding ferrous metals

§59 Safety Devices

The combustion of any fuel gas in an enclosed space always presents a potential explosion hazard. Precautions and devices are necessary to ensure safety. The principal methods of providing protection are:

- a safe, sequential ignition system or procedure
- adequate methods of continuous monitoring of the pilot and the main gas flames during the whole time that gas is supplied
- the provision of approved safety control devices
- the use of flame traps where stoichiometric mixtures of air and gas are conveyed by pipework
- for plant operating at temperatures below the incandescent range, that is below 600°C (1110°F), the provision of approved explosion reliefs.

The design and operation of burner control systems and devices is dealt with in Chapter 4 and flame monitoring systems are discussed in Chapter 5.

Explosion reliefs are required by the Department of Employment to be fitted to plant classified as "ovens". The recommendations specifies the:

- size and design of relief
- materials of construction
- location
- methods of fixing

An explosion relief is a weak section of the oven which will give way at a low pressure and provide an opening for the gases to vent quickly and safely.

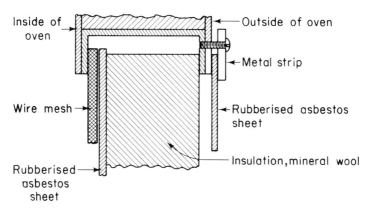

Fig. 24. Explosion relief

Box ovens usually have the relief located at the back of the oven and an approved design is shown in Fig. 24. The oven must be spaced at least 380 mm (15 in) from any wall immediately behind it, and if sited in a corner, at least 600 mm (2 ft) from both side and rear walls. Personnel must be prevented from entering any space into which flames from a relief may be discharged.

Treble cased ovens usually have two reliefs, one in the inner case and one in the outer double case. The two reliefs are at the back of both compartments but do not interconnect.

Conveyor ovens usually have open ends through which some pressure would be dissipated. Ovens where the length is more than six times the diameter may require additional reliefs.

For furnaces operating at temperatures in the incandescent region no explosion reliefs are necessary. Any stoichiometric mixture admitted will burn steadily and not explosively. These furnaces do, however, present other problems.

Many high temperature furnaces work continuously for many months without being shut down. Some would be destroyed if the temperature was to fall suddenly or rapidly. In addition to damaging the furnace, an unexpected shut down could also spoil the contents of the furnace resulting in an entire batch of rejects. Failure of a furnace can be a very costly business indeed.

For this reason extra precautions are taken. Continuous checking ensures that the equipment is operating satisfactorily. Control systems may be duplicated or triplicated. Flame monitors may be arranged so that shutdown will only occur if two out of the three systems simultaneously call for action.

Complex control systems may be expensive but cost of failure would be many times the cost of the protection system.

Furthermore, before carrying out any work on the design, installation or commissioning of industrial process plant, reference should be made to current recommendations. These include:

- Gas Safety Regulations
- British Gas Corporation Publications
- HMSO publication on Health and Safety at Work
- Manufacturer's instructions

CHAPTER 4

Industrial Gas Burner Systems and Their Control

Chapter 4 is based on an original draft by Mr. J. R. Cornforth

BURNERS AND SYSTEMS

§60 Introduction

Gas burners were dealt with in Volume 1, Chapter 3 which concentrated on their principles of operation and their application to domestic appliances. This chapter goes on to consider industrial and commercial burner systems.

Inevitably there is some apparent overlap. Some of the burners used in small commercial and low temperature industrial equipment are very similar to some domestic burners. However, burners used in larger equipment and high temperature applications are very different both in the pressure of the air and gas used and in the manner of their mixing.

The varied requirements of commercial and industrial equipment call for a variety of flames which differ in their four main characteristics:

- shape
- size
- aeration
- temperature

Suitable burners may be selected from a wide range of proprietary types or be developed by the industry for a special application. For any given application there may be more than one way of achieving a satisfactory result.

§61 Types of Burner System

Burners for commercial or industrial equipment may be divided into five main categories as follows:

- diffusion flame or postaerated burners
- atmospheric or natural draught burners
- air blast premix burners
- nozzle mixing burners
- other burners and special applications

Diffusion flame burners, or neat gas burners, obtain all their air for combustion from the surrounding atmosphere.

Natural draught burners use an atmospheric injector assembly, Fig. 1.

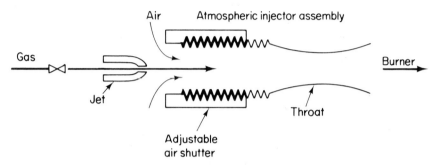

Fig. 1. Natural draught burner

Here, the gas mixes with a proportion of the combustion air before entering the burner. The air, at atmospheric pressure, is entrained by gas under pressure issuing from a jet. If the gas is at normal supply pressure, the burner is a "low pressure" type and the primary aeration is less than stoichiometric. The remaining air is obtained as secondary air from the atmosphere into which the burner is firing. If gas is available at relatively high pressures of about 2 bar (30 lbf/in^2), stoichiometric proportions may be obtained by using a high pressure natural draught burner.

Air blast burners also achieve stoichiometric air/gas mixtures by using air at high pressure to entrain gas at atmospheric pressure. Because all the air required for combustion is available in the burner nozzle, combustion is much more rapid and a short, intense flame results. The system uses fanned air at about 75 mbar (30 in w.g.), Fig. 2. Both this and the natural draught system are known as 'premix' systems because air and gas are mixed in varying degrees before they reach the burner. The same result could be achieved by a mixing machine, but the initial and running costs would be considerably higher.

Nozzle mixing burners are supplied with both air and gas under pressure and there is no prior mixing until they reach the burner nozzle, Fig. 3. The air and gas are proportioned separately by linked

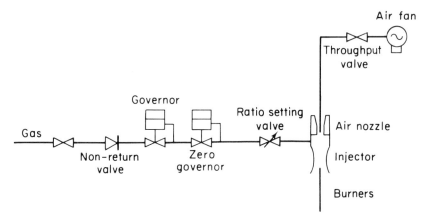

Fig. 2. *Air blast burner system*

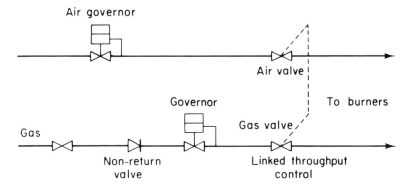

Fig. 3. *Nozzle mixing burner system*

valves or other techniques and fed independently to the burner nozzle. Mixing, by this method, is very positive and mixtures are usually stoichiometric. Intense rates of combustion are developed for high temperature work. The burners are usually very flexible and can be run on either air or gas rich mixtures. They have a greater degree of turn down than a corresponding air blast burner.

Other burners and applications include those which are either outside the four main categories or are special applications of them. They include:

- radiant, or surface combustion burners
- catalytic combustion burners
- radiant tube burners
- self recuperative burners
- packaged burners
- dual fuel burners

§62   Diffusion Burners

The diffusion flame often burns with a luminous appearance. This is due to the unburnt gas issuing from the burner being "cracked" by the heat developed in the burning outer flame. Minute carbon particles are formed which, on reaching the outer zone of the flame, react with oxygen and burn with a yellow luminosity.

Because of the low flame speed and high air requirements of natural gas there are relatively few burners of this type used. Large power station water tube boilers often inject neat gas into the combustion chamber where there is ample space for combustion.

Air recirculating ovens, used for drying processes, often employ a suction burner, frequently the "Firecone" shown in Fig. 4. The recirculating fan creates a negative pressure inside the combustion chamber and combustion air is induced naturally.

Whilst many small industrial jets are now aerated, some small diffusion flame burner heads are produced. Figure 5 shows one in which neat gas issues from two opposite holes and impinges on two stainless steel wings to give a fan shaped flame.

Other neat gas burners include pinhole burners, target burners and matrix burners which are described in Volume 1.

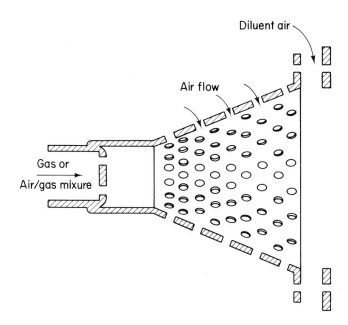

Fig. 4.   Diffusion burner (W.M.G. Firecone)

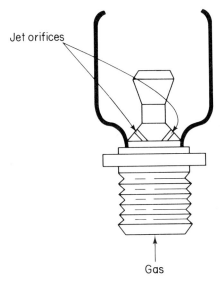

Jet orifices

Gas

Fig. 5.   Neat gas jet (Drew)

§63   Natural Draught Burners

The majority of these are low pressure natural draught burners, using gas at 15 to 20 mbar (6 to 8 in w.g.). The energy of the gas is used to entrain about 50 to 55% of the air required for combustion in the form of primary air. The burners usually operate in open atmospheric conditions.

Uses

Burners of this type form the largest proportion of industrial burners. They are used for relatively low temperature processes in the range of 200 to 400°C (390 to 750°F). Applications include small box ovens and underfired liquid heating tanks.

Types

Probably the most common type is the simple drilled bar burner. Fig. 6. Other examples in general use range from simple single port burners, such as the bunsen burner, to large assemblies of drilled or ribbon–ported bar or ring burners.

The burner port area for a particular heat input rate is termed the "burner port loading". It may be calculated from:

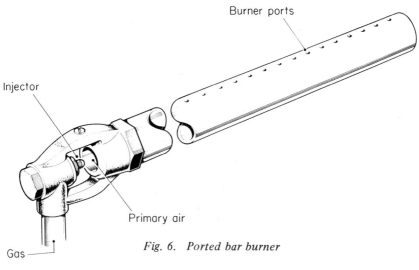

Fig. 6. *Ported bar burner*

$$\begin{array}{l} \text{burner port loading} \\ \text{MW/m}^2 \\ \text{(Btu/in}^2\text{h)} \end{array} = \frac{\text{heat input rate, MW (Btu/h)}}{\text{total burner port area, m}^2 \text{ (in}^2\text{)}}$$

For drilled ring burners with about 50% primary aeration, burner port loadings should be limited to about $10\,\text{MW/m}^2$ ($20\,000\,\text{Btu/in}^2\text{h}$), or lift off will occur. For lower primary aeration, port loadings may be increased to about $25\,\text{MW/m}^2$ ($50\,000\,\text{Btu/in}^2\text{h}$), but care must be taken to avoid yellow tipping and soot formation.

Flame Retention

If higher burner port loadings are required, some method of flame retention is necessary. With natural draught burners there are three main types:

- sudden enlargement of mixture flow
- wake eddies produced downstream of an obstacle
- stabilisation by auxiliary flames

Sudden enlargement is achieved by welding or rivetting continuous strips along the bar on either side of the main flame ports, (Fig. 7). This gives a sudden enlargement of the mixture flow cross section and generates a recirculation of hot gas products towards the root of the flame, so providing a source of heat for ignition. The strips also shield the flames and reduce the effect of cross draughts. They should be saw cut at intervals along their length to reduce any bending effect caused by unequal expansion of the strips and the bar material.

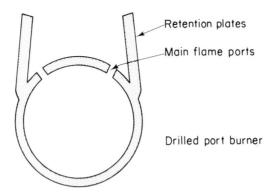

Fig. 7.   Flame retention; sudden enlargement

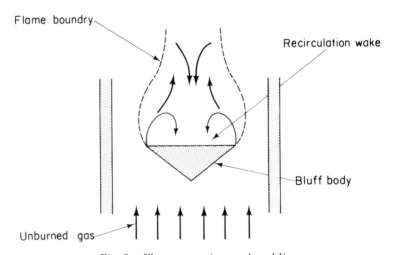

Fig. 8.   Flame retention; wake eddies

Wake eddies have the effect of producing a zone in which mixture velocity is reduced, giving a higher probability of the mixture becoming heated for ignition. They also create a recirculation behind the eddy, Fig. 8.

Auxiliary flames or retention flames are, in effect, pilot flames, generated by much smaller ports adjacent to the main flame ports. A flame retention plate is welded or rivetted along each side of the main flame ports, Fig. 9. The small pilot drillings act as metering orifices and slow down the rate of mixture flow so that a continuous strip of pilot flame forms between the bar and the retention plate. This flame has the effect of constantly keeping the main flame lit. With this design there may be problems due to the retention strip buckling or dust deposits collecting in the metering orifices.

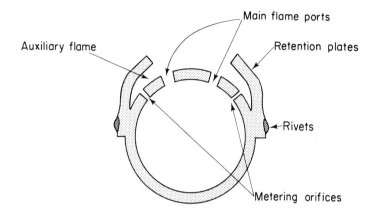

Fig. 9.   Flame retention: auxiliary flames

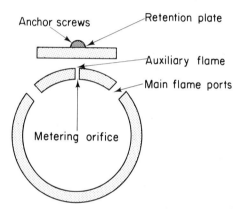

Fig. 10.   Horizontal flame retention plate

A form of this burner which overcomes both these problems is shown in Fig. 10. The retention plate is secured in a horizontal position by screws in elongated holes which allow for expansion. The plate prevents dirt from falling into the central metering orifices. The burner gives short well defined flames at high outputs and is completely stable. This is because secondary air is easily entrained by the roots of the main flame.

The ribbon burner, Fig. 11, also produces auxiliary flames. The ribbon forms a combination of large and small burner ports, the small ports being located at either side of each crest of the corrugations. These burners are very versatile and can be used with high and low pressure natural draught injectors. They may also be used for air blast and premix systems giving stoichiometric aeration. The

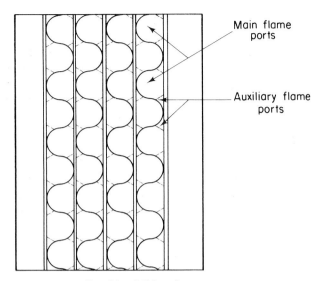

Main flame ports

Auxiliary flame ports

Fig. 11. Ribbon burner

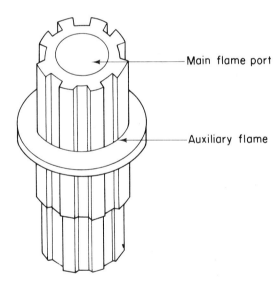

Main flame port

Auxiliary flame

Fig. 12. Aeromatic "Insertajet"

ribbon must be securely welded or rivetted in position otherwise it may be forced out when the burner gets hot.

A simple method of providing auxiliary flames in fairly high rated bar or ring burners is by the use of the "Insertajet" (Aeromatic Limited), Fig. 12. Originally produced for conversion, these jets are now available in larger outputs up to 300 W (1000 Btu/h) per jet at normal gas pressures. They may also be used at higher pressures. The burner ports are drilled out to 5.9 mm ($\frac{15}{64}$ in) diameter and the "Insertajet" tapped lightly in until it reaches the shoulder. The jets must be a tight, interference fit otherwise they may come loose when the burner heats up. On ring burners it is only necessary for 50% of the ports to be fitted with "Insertajets". This is adequate to ensure that the remainder of the flames are continuously lit.

Pilot Burners

When fanned air is not available, natural draught pilots are often used for smaller burner installations. They usually have an adjustable air port to vary the hardness of the flame and they can be associated with either thermoelectric or flame rectification monitoring systems, Fig. 13. The primary function of the pilot burner is always to keep the main burner alight.

Where more than one bar burner is fed from a single control valve, a "ladder pilot" may be fitted. This is a small bar burner with ports

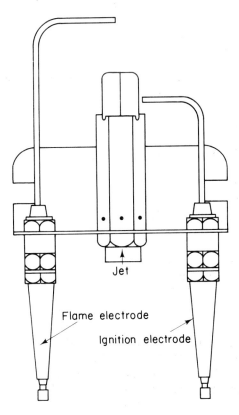

Jet

Flame electrode

Ignition electrode

Fig. 13.   Natural draught pilot burner

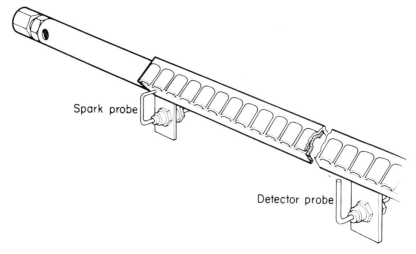

Fig. 14.   Ladder pilot burner

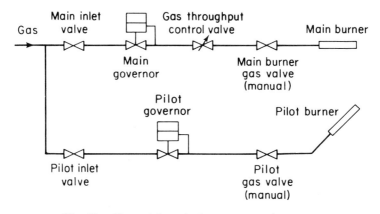

Fig. 15.   Natural draught burner control system

that give a continuous pilot flame from one end to the other, Fig. 14. The ladder pilot is fitted at right angles to the main burners and the flame sensor must be at the opposite end to the ignition device. This should ensure that all main burners are alight when the flame is detected (see Chapter 5).

Control Systems (non-automatic burners)

Because the gas throughput on natural draught systems is generally relatively small the controls are kept simple. A typical control train is shown in Fig. 15. The pilot line is separately valved and governed so that the pilot is not starved of gas when the main valve is opened.

To ensure safety, a flame protection device should be added (see Chapter 5). Where the heat input rate is not more than 120 kW (400 000 Btu/h), for example in a drilled bar burner, thermo electric flame protection is considered adequate (provided that the closing time following flame failure is not more than 45 sec). For rates above this, either flame rectification or ultra-violet devices should be used in conjunction with at least one certificated safety shut off valve.

§64   High Pressure Natural Draught Burners

Where the gas supply is available at relatively high pressures, a natural draught injector may be used to provide near stoichiometric mixtures. For example, a gas pressure of 2 bar (30 lbf/in^2) will give a stoichiometric mixture pressure of 6.25 mbar (2.5 in w.g.). This is sufficient pressure for the burner to be stable over a satisfactory turndown range. Alternatively, if this pressure is not available, a typical natural draught system could be employed with the remaining air being supplied as secondary air. If higher gas pressures are required a reciprocating or vane type booster could be used. Air blast systems can offer similar advantages with much simpler equipment.

§65   Mechanical Mixing Techniques

In some installations natural draught burners may be supplied from a mixing machine. These systems were often used during conversion to reduce the cost. By producing an air/gas mixture of the same Wobbe Number as manufactured gas, the existing burners could still be used. In some cases the whole factory may be supplied by one mixing machine, in others it could be a single item of plant, for example a continuous baker's oven with a hundred or so ribbon burners.

In most mechanical premixing machines, the air and gas are drawn through suitable metering elements into a mixing device and then compressed. The rotary compressor is the most widely used type. It produces a mixture of substantially constant air/gas ratio at pressures up to about 350 mbar (5 lbf/in^2). It is important that the volume of air inspired should form a gas rich, non-explosive mixture which may then be safely distributed to a number of burners with atmospheric injectors to give a final mixture near to stoichiometric. A simplified system is shown in Fig. 16. An essential feature of the system is the ability of the machine to deliver a constant ratio mixture when the back pressure or the number of burner heads in use is varied.

For safety it is essential to fit a low pressure cut off device in the gas supply to the compressor. Additionally, a flame trap element

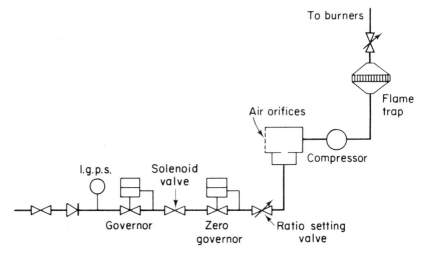

Fig. 16. Simplified layout of mechanical premix machine

should be fitted in the mixture supply as near to the burners as is practicable.

The system described gives only a "partial premix" requiring some atmospheric air to complete the stoichiometric proportions. A "full premix" system can be used to provide a stoichiometric mixture. In this case the most stringent safety measures must be taken, particularly with regard to the use of flame traps which may need to incorporate fusible links or over-heat cut offs.

§66   Air Blast Burners

This system uses air under pressure to entrain gas at atmospheric pressure, the opposite of the natural draught system. Because the air pressure is commonly about 70 mbar (1 lbf/in^2) a stoichiometric mixture is easily produced. As all the air required for combustion is delivered to the burner head, this results in intense combustion and a fairly short flame.

Uses

Air blast burners are generally used for higher temperature work than diffusion flame burners where high heat inputs are required. They are frequently used on all types of heat treatment and annealing furnaces and also on furnaces for reheating steel prior to forging small articles. With correct furnace design, temperatures up to 1200°C (2190°F) can easily be achieved.

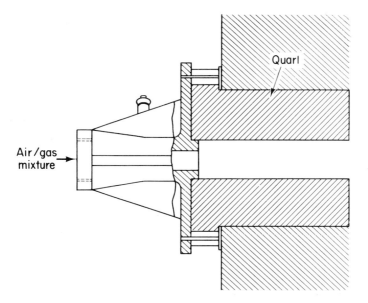

Fig. 17.   Air blast burner with integral quarl

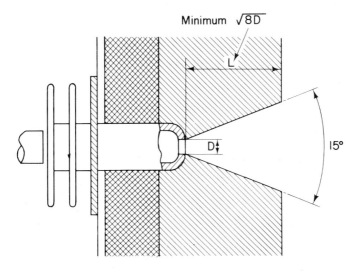

Fig. 18.   Air blast burner firing into flared tunnel

Types

These burners may be divided into two distinct groups:

- burners firing into a refractory "quarl"
- burners without refractory quarls

The 'quarl' is a refractory burner head or tunnel. Where this is integral with the burner casting it may be sealed into the furnace wall, Fig. 17. In other types the cast iron burners are bolted on to the side of the furnace casing and a suitably flared or parallel tunnel is then cut in the furnace refractory wall, Fig. 18. In these types of burner the flames are stabilised by radiation from the incandescent quarl. It is therefore important that quarls are correctly designed. For a parallel tunnel the relationship between the length and the diameter are as follows:

If the diameter of the air jet nozzle is d
And the diameter of the mixture tube orifice is D
Then:

$$D = 1.6 d$$

Tunnel diameter $= 4.8 d$
Tunnel length $= 16.0 d$

Flame retention

When air blast burners are used without refractory quarls, a flame retention head or non-blow-off burner tip is required. Small auxiliary flame ports are incorporated into the burner head to produce low velocity pilot flames which keep the base of the main flame alight. A typical burner tip is shown in Fig. 19. It has small metering orifices

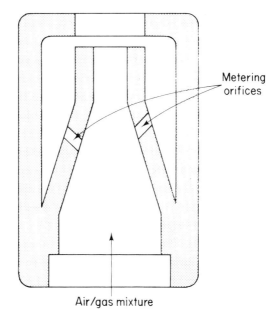

Metering orifices

Air/gas mixture

Fig. 19. Non-blow-off tip

drilled around the inner core. The air/gas mixture then expands between the concentric tubes and forms a low velocity stable flame which keeps the root of the main flame continuously lit. The tip allows the use of high mixture velocities without blow-off.

These tips are usually made from steel and designed for use in the open, to avoid over-heating and to prolong their working life. They may also be used for "gap-firing". Here the tips are spaced at a short distance from the furnace wall and fire into a circular port in the wall refractory. With this method, excess atmospheric air is induced into the air/gas mixture stream and it is used where lower furnace temperatures are required. Gap-firing also keeps the tip relatively cool.

Control Systems

In a typical system air from a fan at about 75 mbar ($1 \, lbf/in^2$) is allowed to expand through the jet of an air blast injector into a venturi mixing tube, Fig. 20. The expanding jet of air entrains gas available at zero, or atmospheric pressure, which is admitted to the mixture tube immediately downstream of the air nozzle. The mixture is then supplied directly to the burner. There must be no restriction or valve between the injector and the burner. Providing that the gas is supplied at zero pressure, usually by means of a zero governor, and that the burner is firing into approximately atmospheric pressure, then the quantity of gas entrained is proportional to the air flow through the injector. The gas ratio setting valve is often

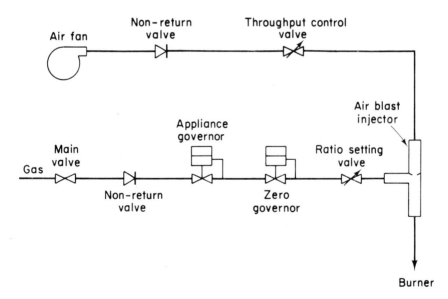

Fig. 20. Air blast control system

incorporated into the injector, when it is known as the "obturator setting screw".

In the tunnel burner, injection, mixing and combustion occur together in the tunnel. With this and any other system using fanned air, it is a requirement of the Gas Act that a suitable non-return valve be fitted in the gas supply line. This is to prevent any air from entering the gas distribution system. It is also advisable to fit a non-return valve in the air supply line, especially if the fan is situated above the burner, to prevent gas entering the fan housing and escaping to atmosphere.

Commissioning

The throughput control valve, which is a single linear flow type valve or quadrant cock in the air supply line, should be set to "low fire" rate and the burner ignited. The air/gas ratio should be set, while at low fire rate, by adjusting the zero governor tension spring. The gas pressure downstream of the governor should then be substantially at zero gauge pressure.

The system should be turned to high fire rate, by the quadrant cock and the air/gas ratio set by adjusting the obturator screw or a separate ratio setting valve. It may then be necessary to re-check the setting at low fire rate. When set correctly proportional control is achieved by the one valve in the air supply line.

The setting up of the system is made easier if flow meters can be fitted in the air and gas lines or if an orifice plate or insertion meter can be used. In most cases, on the district, the system must be checked by viewing the burner; the correct settings may be determined with experience. Flue gas analyses should be used whenever possible.

Appliance or Mixture Pressure Back Loading

If the chamber into which the burner fires is not at atmospheric pressure, then to maintain the self proportioning action, the reference pressure for both the appliance and the zero governors should be the appliance chamber pressure and not atmospheric pressure.

To compensate for changes in appliance back pressure, the top diaphragms of both the appliance governor and the zero governor are back loaded with the furnace chamber pressure. This is usually communicated by small bore copper tube from a position in the appliance adjacent to the burner. Furnace chamber pressures are seldom higher than 5 mbar (2 in w.g.).

Compensation for changes in mixture pressure is obtained by mixture pressure back loading the two governors from a position in the mixture manifold. Apart from the point of connection, the two

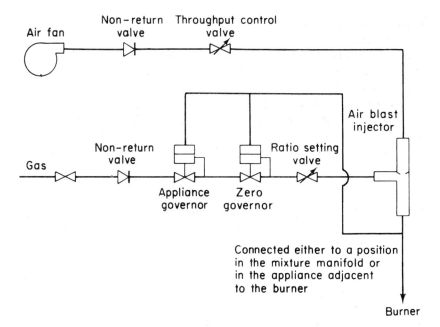

Fig. 21.   Appliance or mixture pressure back loading

systems are identical, and are shown in Fig. 21. With any back loading technique it is necessary to have the inlet pressure to the zero governor at least 0.75 mbar (0.3 in w.g.) higher than the back loading pressure and the top chambers of both governors air tight.

§67   Nozzle Mixing Burners

In this system, gas and air do not mix until they enter the burner quarl. The burner body merely serves as a distribution box, conveying the air and gas to the quarl. The other difference between this and the previous systems is that both air and gas are under pressure. The burners are generally very versatile. They have a wider turn down range than the equivalent air blast burner and can be used with either excess air or excess gas. Because all the air for combustion is supplied to the burner, relatively rapid mixing results giving maximum heat release rates and short flames.

Uses

Nozzle mixing burners are used on all types of high temperature plant, from small heat treatment furnaces to large reheating furnaces and non-ferrous metal melting furnaces. Because of their flexibility and relative ease of commissioning, these burners are generally replacing the equivalent air blast systems.

Types

There are many different types of nozzle mixing burners. In each case, flame retention is achieved by the recirculation of hot products in the quarl and also by radiation from its incandescent refractory walls. This enables the burner to operate satisfactorily at low through-puts, both in stoichiometric conditions and with large amounts of excess air. If a more gentle flame is required for less intense local heating, gas rich firing can be employed.

A typical burner is shown in Fig. 22. Here, both the gas and air flows are axial, with the air annulus surrounding the gas flow. In some designs a small amount of air may be fed into the gas stream before the quarl either axially or tangentially, to assist flame retention.

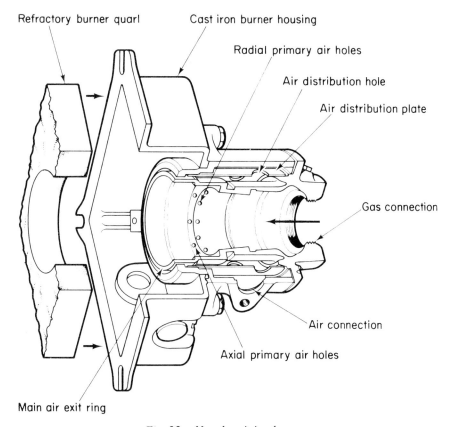

Fig. 22. Nozzle mixing burner

To give even more rapid rates of heat release some burners are fitted with swirl vanes in the gas and air sections to increase the speed of mixing.

High velocity burners are used for processes which rely mainly on forced convection as the method of heat transfer. In this case the main obstacle to efficiency is the thin film of gas on the surface of the stock. Increasing the velocity of the heating gases relieves or removes the insulating film so increasing the heat transfer efficiency. A high mass flow of hot gases, forced to envelope the stock completely, results in high speed heating with very low temperature differentials. Because there are no high temperature gradients, hot spots cannot form and overall uniformity of temperature is improved.

This can be achieved by using recirculating fans or, more simply, by utilising the jet effect of a high velocity burner, Fig. 23. Because combustion is completed within the burner and because the exit quarl is convergent, the products leave the burner with a high velocity. In a properly designed furnace the thrust from the burner can be used to entrain combustion products and recirculate them around the load. The burner can be operated on high excess air so it is equally suitable for low as well as high temperature applications.

Several small burners in a furnace could be replaced by one single high velocity burner, so giving even temperature distribution and making the provision of flame protection much simpler.

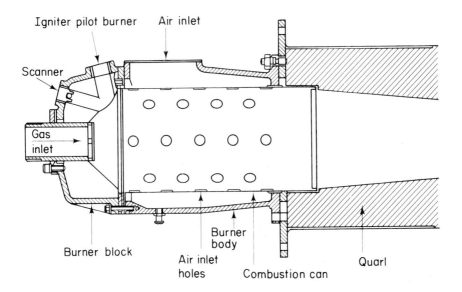

Fig. 23. High velocity burner

Flat flame burners are used when flame impingement on the stock must be avoided. A typical burner is shown in Fig. 24. It is essential that the outlet of the refractory quarl is in alignment with the furnace wall or roof so that the flame can flow from the burner on to the adjacent surface. The air inlet orifice is arranged to swirl the air so that the combustion products tend to hug the surface of the quarl and move along the surface of the furnace wall with little forward velocity. There is, therefore, no flame impingement on the stock. The successful uses of this burner include galvanising bath heating and crown firing of small beehive brick kilns.

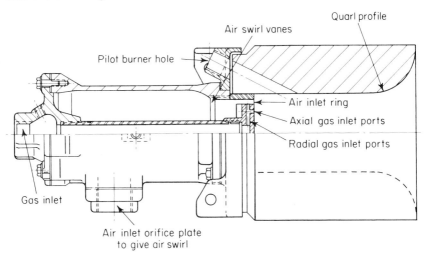

Fig. 24.   Flat flame burner

Control Systems

The two main systems for the control of nozzle mixing burners are:

- the linked valve system
- the pressure divider technique

Linked valves control the separate flows of air and gas to the burner. One valve is situated in the air supply line and the other in the gas supply. They are linked either mechanically or electrically so that the flows of air and gas can be adjusted simultaneously. The mechanical linkage is usually driven by an electric motor and the electrical link is usually operated electrically or pneumatically. In some simple systems the two valves are adjusted independently of each other, especially if the air/gas ratio is required to vary during the process cycle.

Changes in back pressure are often appreciable compared with the

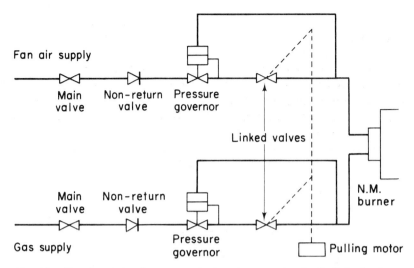

Fig. 25.   Nozzle mixing burner control system, mechanically linked valves

supply pressures and tend to affect the air and gas flows unequally, so causing a deviation from the ratio required. Similarly, if pre-heated air is used, the air/gas ratio can vary from the desired value due to the change in air density at the burner nozzle. If, however, the differential pressure across the valves is kept substantially constant, a suitable choice of valves can maintain the air/gas ratio to within narrow limits. Control of the differential pressure is achieved by fitting pressure governors up stream of the valves, back loaded from pressure tappings down-stream, Fig. 25. By positioning the air valve up-stream of any recuperator, the air controlled by the valve is virtually at constant temperature and density.

If the changes in back pressure are relatively small, adequate control may be achieved without back loading. This is done by using constant up-stream supply pressures which are large in comparison to the pressures down-stream of the control valves.

The valves most commonly used for high-low control are linked butterfly valves. With these it is relatively easy to set the flows to the correct air/gas ratio at both high and low limits. Variations in the ratio may occur at mid-positions during the changes between high and low limits. Care must be taken to ensure that these variations do not seriously affect the stability of the burner.

Another system frequently used for high/low control is the weep relay valve system, Fig. 26. This has solenoid valves situated in the weep lines of relay valves in the gas and air supplies. The solenoids are controlled by a thermostat or similar device. The flow rates at high and low positions are set by adjustable stops above the diaphragm and below the valve of the relay valve.

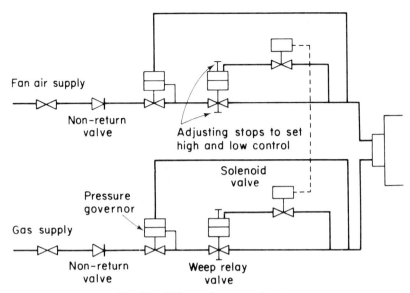

Fig. 26.   Relay valve control system

If proportional or modulating control is required it is essential to match the air and gas flows over the whole throughput range. This requires linear flow valves such as the adjustable port valve. The installation would be as in Fig. 25.

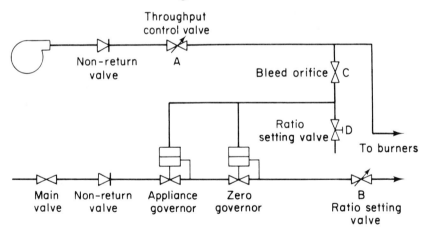

Fig. 27.   Pressure divider control system

The Pressure Divider Technique is applicable when there is little or no injection or pressure inter-action between gas and air streams in the burner. This situation occurs, for example, when a simple concentric tube burner fires into a large chamber. In this case the burner consists of air and gas orifices with the two streams mixing at a common down-stream pressure. Consequently, if the pressures up-stream of the burner are held at a constant ratio over the whole throughput range, a constant air/gas ratio will be maintained. The control lay out is shown in Fig. 27.

A zero governor is fitted in the gas supply and is loaded with a fraction of the air supply pressure from pressure dividing orifices. These are a bleed orifice C and a ratio setting valve D. The system has the advantage that throughput may be controlled by a single valve, A in the air supply line. As the air flow through the valve is increased, the diaphragm of the zero governor is further depressed by the bleed air and the gas flow increases in porportion. The bleed orifice and the ratio setting valve are usually incorporated in one proprietary valve body.

When the gas pressure is less than the air pressure, valve B is not fitted.

Commissioning Pressure Divider Systems

Set the valve A to low fire rate and adjust the zero governor tension spring to give a down-stream pressure just above atmospheric, con-sistent with a satisfactory flame at the burner.

Turn to high fire rate and adjust by trimming the ratio setting valve D, provided that the correct size bleed orifice C has been fitted. It may then be necessary to readjust the low fire setting. When the system has been set up all variable adjustments should be locked in position. Once correctly set the system will give a constant air/gas ratio over the whole turn down range so, if a linear throughput valve is used, the system will give proportional control.

In an installation in which the gas pressure is equal to or greater than the air pressure, the bleed orifice is removed so that full air pressure is fed to the top of the governor diaphragms. The ratio setting valve is located at B and the pipework at D is blanked off. Low fire is adjusted by means of the zero governor tension spring and high fire rate is adjusted by the ratio setting valve B.

OTHER BURNERS

§68   Radiant Burners

These burners were introduced in Volume 1, Chapter 3 and some types are used on domestic appliances. The burner heads are of porous or perforated refractory material or of wire gauze. They are known as "surface combustion" burners because combustion usually takes place at the surface of the porous medium or about 1.5 mm below the surface in the perforated medium. The air/gas mixture is supplied either from an air-blast injector or an atmospheric injector and produces an area of even radiation as the surface becomes incandescent. Surface temperatures of about 850°C (1560°F) are common although special burners can reach over 1400°C (2550°F).

Uses

Radiant burners have been extensively used in overhead radiant space heaters. They are also used for special, high temperature drying operations and as the heat source in large, overfired grills.

Types

Burners may be divided into categories based on the type of radiating surface, as follows:

- porous refractory
- perforated refractory
- wire gauzes
- catalytic combustion
- high-temperature porous medium burners

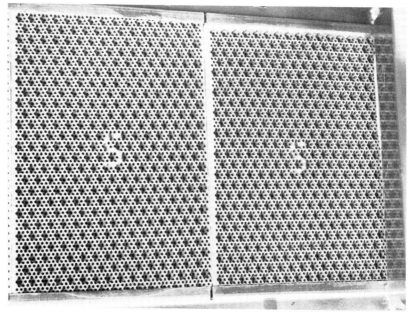

Fig. 28.   Perforated refractory burner (Schwank)

Porous refractory burners may have either atmospheric or air blast injectors producing an air/gas mixture which is fed through the pores in a refractory plaque.

Perforated refractory tiles are used in burners like the Schwank burner, Fig. 28. This has perforations about 1.4 mm (0.055 in) diameter set in individual plaques of about 65 × 45 mm (2.6 in × 1.8 in) built up to form the radiating area. The air/gas mixture is supplied from an atmospheric injector.

Wire gauzes are used in some domestic grills and in some over-head radiant heaters (Chapter 9). They are suitable for low temperature applications. A metal gauze is fitted about 5 mm (0.2 in) above the plaque surface on perforated refractory burners when burning natural gas, to give flame retention.

Catalytic combustion has been widely used on the Continent, principally with lpg. The gas reacts with a catalyst to burn at a low temperature, about 450°C (840°F), without an apparent flame. Work on developing a more effective catalyst for natural gas is continuing.

High temperature porous medium burners have been developed by research laboratories and the manufacturers to give operating tem-

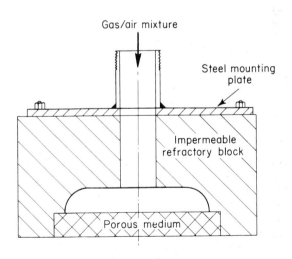

Fig. 29.   Porous radiant burner (Shell)

peratures up to 1400°C (2550°F) with high radiant efficiency. The Shell P.R. (Porous Radiant) Burner, Fig. 29, using porous elements made either from:

- zircon, or
- sillimanite

The zircon burner can be operated at higher surface temperatures than the sillimanite, but the sillimanite panels offer less resistance to the mixture flow and can be operated on lower mixture pressures.

§69   Radiant Tube Burners

These were introduced in Chapter 3 and are used where indirect firing is essential. For example, in many heat treatment processes where the work must not come into contact with the products of combustion or the flame.

The tubes can provide a range of temperatures from about 300 to 1100°C (570 to 2010°F). They may be used for a large number of industrial processes. Radiant tubes can be divided into three main categories:

- straight-through
- double pass single-ended
- recirculating

Straight-through types include tubes of the following shapes:

- straight
- curved parabolic
- 'U' shaped
- 'W' shaped

Each of these tubes may be fitted with external recuperators, if required.

Double pass single-ended tubes have the burner and flue combined at one end. They have an internal tube which transmits the hot gases to the closed end of the outer tube where they are diverted back through the annulus in a counter-flow direction. A refinement of this principle incorporates a recuperator and Fig. 30 shows a burner of this type. Thermal efficiency may be about 60 to 70%. Single ended tubes may easily be replaced in a furnace, unlike the 'U' and 'W' shaped tubes.

Recirculating types of radiant tubes have a concentric inner section which further increases the operating efficiency of the tubes by recirculating the hot combustion products, as in the M.R.S. developed burner. A ceramic tube, of silicon carbide bonded with silicon nitride,

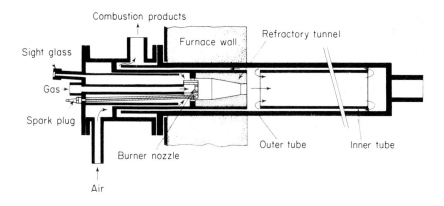

Fig. 30. Radiant tube burner with recuperator

of the same type will withstand process temperatures up to 1250°C (2280°F) and is suitable for bright annealing of stainless steel and for special ceramic wares.

§70 Self Recuperative Burners

Waste heat recovery can be achieved either by an external heat exchanger or by a recuperator integral with the burner. The latter system avoids the heat losses experienced from the pipework associated with external recuperators. There is, however, the problem of air density variation as the burner heats up.

A typical burner was described in Chapter 3. It consists essentially of a high velocity nozzle mixing burner surrounded by a counter flow heat exchanger supplying hot combustion air to the burner nozzle. This can give air preheating up to about 650°C (1220°F).

Normally, all the combustion products are extracted through the recuperator by means of an air driven eductor mounted on the burner flue outlet. The furnace pressure is maintained constant by regulating the eductor air supply.

The burners may be used for a variety of processes including batch furnaces for heat treatment of metals, non-ferrous metal melting, salt baths and pottery kilns.

§71 Packaged Burners

These are self contained units for use on industrial or commercial appliances. They are easily installed by bolting on to the combustion chamber of the appliance and connecting to gas and electrical supplies. If the rating of these burners is between 60 kW (205 000 Btu/h) and 2 MW (6.8 million Btu/h) they should be designed to comply with British Gas Standard for Automatic Gas Burners, Forced and Mechanically Induced Draught (Publication IM/8).

Packaged burners normally consist of:

- burner head
- combustion air fan
- programming control unit, with ignition and flame detection devices
- two safety shut off valves
- main gas governor, and pilot gas governor
- pilot solenoid valves
- air proving device
- isolating valves

The programming control unit and the safety shut-off valves should be of types certificated by BGC Midlands Research Station.

Sequence of Operation for Burners complying with British Gas Standard for Automatic Gas Burners, Forced and Induced Draught

This is monitored by the programming control unit and consists of the following operations:

1. A pre-purge period of proved air pressure of at least 30 second, at the full combustion air rate. Five air volume changes must be given.

2. A safe start check. Flame simulation in the pre-purge period must cause lockout.

3. A start gas flame ignition period of no longer than 5 seconds. This is the time when the spark and the pilot gas valve are both energised, and when no check is made for the presence of the flame.

4. A start gas flame proving period of at least 5 seconds. This follows immediately upon the previous period, and the pilot flame must be sensed throughout this period. During this time the pilot flame is alight without the spark.

5. The main flame ignition period, of no longer than 5 seconds. This is the time when the pilot burner should be lighting the main burner, and therefore main and pilot valves are open. Again, the flame must be sensed throughout this period.

6. The main burner run period. The pilot valve is closed at the end of the previous period, and hence the main burner is

alight alone. The flame must be sensed throughout this period. In modulating systems, this period may be split into the main flame proving period, after which the burner is switched to full modulation control.

7. A post purge period. This is optional, and comes into operation on shut-down or lock-out of the burner.

8. If the flame is lost in either the pilot flame proving or the run position lockout will occur. When lockout does occur, this will require manual intervention by an operator to reset the unit to restart the system.

9. With modulating units:—

   • the air flow damper for pre-purge must be proved in the high fire position
   • the air and gas throughput dampers are proved in the low fire position at start up only

10. Loss of air at any time must give either shutdown or lockout. The air proving device must be proved in the no-air position prior to start up, otherwise start up should be prevented or lockout result.

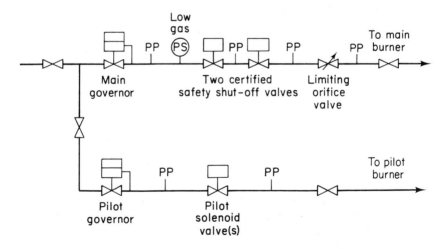

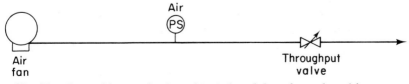

Fig. 31. Control layout for forced or induced draught packaged burner

The flame safeguard should be checked at least daily on continuously operated plant by manually shutting the burner down and allowing it to start up again.

The main safety shut off valve requirements are based on heat input rates as follows:—

- from 60 to 600 kW (205 000 up to 2.05×10^6 Btu/h)—one Class 1 and one Class 2 valves
- above 600 up to 1 000 kW (above 2.05×10^6 up to 3.4×10^6 Btu/h)—two Class 1 valves
- above 1 000 up to 2 000 kW (above 3.4×10^6 up to 6.8×10^6 Btu/h)—two Class 1 valves with a system check

A typical control layout for a packaged burner is shown in Fig. 31.

Uses

Package burners are frequently found on small sectional boilers, commercial air heaters and on a multitude of small, low temperature heating applications.

§ 72   **Large Gas and Dual Fuel Burners**

When the fuel load requirements of an organisation are large, the customer may have an interruptible tarriff agreement. This enables him to purchase gas at an advantageous rate, provided that his plant can operate on an alternative fuel (usually oil), when gas is in great demand, for example, during the winter months.

Dual fuel burners, capable of operating on either gas or oil and large gas burners should comply with the requirements of the Code of Practice for Large Gas and Dual Fuel Burners (Publication IM/7) if the plant has a heat input above 2 MW (6.8×10^6 Btu/h), where any individual burner is rated above 500 kW (1.7×10^6 Btu/h).

Large Gas Burners

For burners fuelled only by gas, the Code is basically an extension of the Standards for Automatic Burners (IM/8) with additional requirements. These requirements for burners over 2 MW (6.8×10^6 Btu/h) rating are as follows:—

1. Each gas supply to a burner or group of burners shall be under the control of an automatic safety shut-off system. For main burners this may be either:

   - double block and vent valves with position checking using valves with mechanical overtravel, or
   - double block valves with a pressure proving system

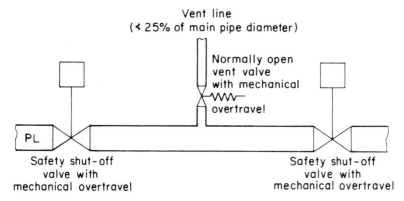

Fig. 32.   Double block and vent system

The first system uses two certificated normally closed safety shut-off valves to close the fuel line and a third normally open valve to vent the space between them to atmosphere. On opening the safety shut-off system the vent valve will be proved closed before the block valves are energised. On closure all valves are de-energised simultaneously, Fig. 32.

Valves with overtravel characteristics are plug, ball or some types of gate valve. They are fitted with proof of closure switches which will initiate lock out if the valves are not in the correct position.

Before the burner start sequence can commence the two normally closed block valves must be proved closed and the normally open vent valve must be proved open.

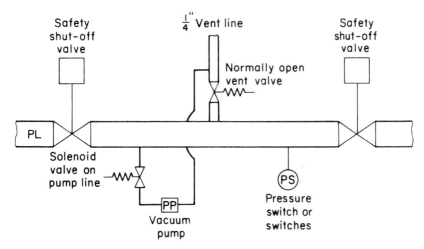

PP < PS < atmospheric pressure

Fig. 33.   Vacuum proving system

The vent pipe bore should not be less than 25% of the main pipe diameter or 15 mm ($\frac{1}{2}$ in), whichever is the greater.

The second system checks the valves for leakage so mechanical overtravel with position proving switches is not required. The systems in common use include:

- vacuum proving
- pressurising with an inert gas such as nitrogen
- opening and closing the valves in sequence using line gas proving pressure

In the vacuum proving system Fig. 33, the space between the two block valves and the normally open vent valve which is made to close, is evacuated by a vacuum pump. The required vacuum should be reached in a set time interval. This checks for a large valve leak. The pump is stopped and a proportion of the vacuum must be maintained over a second time interval. This checks for small valve leaks.

When pressurising with an inert gas the nitrogen, for example, is admitted into the space between the two block valves and the normally open vent valve, which has been made to close. A set pressure must be achieved in a given time interval to prove that the valves are reasonably tight, Fig. 34.

The sequential proving system first closes the vent valve and monitors the pressure in the space between the three valves over a set time interval. If the pressure rises above atmospheric, the upstream

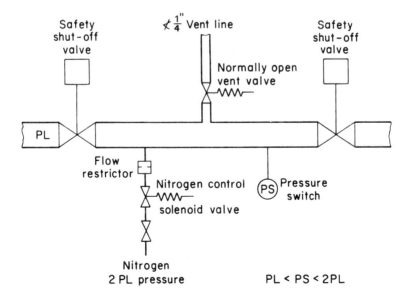

Fig. 34. Nitrogen proving system

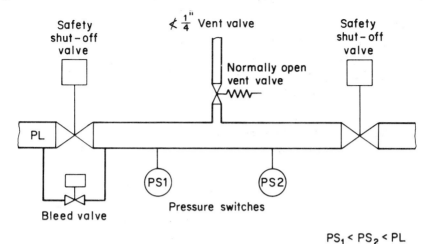

Fig. 35. *Sequential pressure proving system*

valve is leaking and the system will lock out. If this part of the check is satisfactory the upstream block valve (or a bypass) is opened and then closed. This admits line gas pressure into the space between the three valves. The pressure is monitored for a set time interval and, if it drops substantially it indicates that the downstream valve, the vent valve or the valve flanges are leaking, Fig. 35. Vent valves on pressure proving systems should have a port diameter of not less than 6 mm ($\frac{1}{4}$ in).

2. The whole of the pre-purge period must be proved to be in the high air flow condition. This is monitored by a position switch on the air damper or by an air pressure switch down stream of the air damper.

3. Before the main flame establishing sequence can begin, the air and gas dampers must be proved to be in the low fire position. This is done either by damper position proving switches or by air and gas pressure switches.

4. High gas pressure protection is required on systems above 3 MW (10.2×10^6 Btu/h), or where the plant governor inlet pressure exceeds 75 mbar (30 in w.g.), where the pressure drop across the governor is more than 30% of the normal minimum operating outlet pressure.

   Low gas pressure protection is only required if the burner is unstable under lean gas conditions.

5. It is recommended that self checking flame safeguards be fitted. Where these are not incorporated the burner shall be

shut down at least once per day to check the operation of the detector.

Dual Fuel Burners

The burners are principally constructed of components common to both gas and oil systems. These common items are:-

- air fan and air manifold
- programming control unit
- ignition transformer and spark plug
- flame detection system
- pilot burner

The burner heads and fuel control systems are different. Gas burners are usually nozzle mixing with a cast iron pepper pot gas manifold.

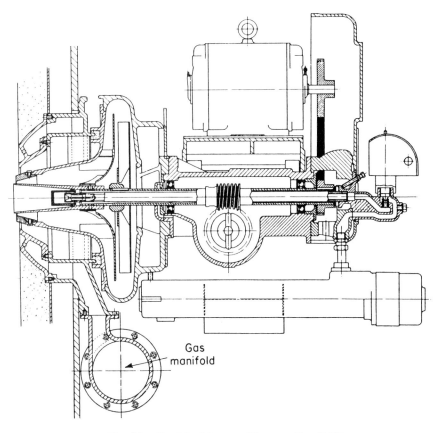

Gas manifold

Fig. 36.   Dual fuel burner (Hamworthy AWI)

The oil burner is usually a spinning cup to accept 950 seconds oil, Fig. 36. Pressure jet burners are used when the stand by fuel is light oil of 28 or 35 seconds viscosity.

The stand by oil system should be adequate for at least one month running. Light oil is generally used but heavy oil may be found if it was in use before gas firing was introduced.

When the system is switched from one fuel to another, one of the following sequences is employed.

1. *Full restart switchover.* This entails the complete interruption of firing on the one fuel and the full start up sequence on the second fuel, including the pre-purge.

2. *Piloted switchover.* This requires establishment of the pilot, the interruption of the main flame firing and the restart of the main flame on the second fuel for each burner in turn. Purging is dispensed with because of proved continuity of the flame.

3. *Dual valve switchover.* This system uses a pair of linked valves, one in each fuel line, separate from the main throughput control valves. The valves are arranged so that, as one opens, the other closes. Both valves are set at a fixed plant load to give a constant total thermal input.

4. *On load switchover.* Switchover is carried out while maintaining the plant under full process control. This also involves the simultaneous reduction of one fuel input rate with a corresponding increase in the other fuel rate while maintaining the total required thermal input.

5. *Sequential shut down switchover.* This entails the complete interruption of firing on one fuel and restarting on the second fuel, excluding the purge, for each burner in turn. The system should only be used on multi-burner plant.

A typical control train for large burners which complies with the Code is shown in Fig. 37.

Uses

The most popular use for these burners is on large boilers, for example, Economic, Lancashire and water tube boilers. They are also applied where large heat input rates are required on low and medium temperature plant.

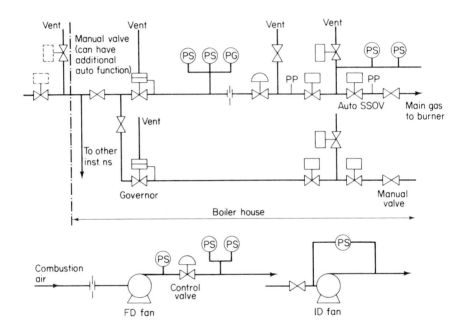

Fig. 37.   Control layout for burner systems over 2MW

CONTROL DEVICES

§73 Introduction

To ensure safe and efficient combustion, gas and air supplies must be adequately and accurately controlled. The devices used on industrial and commercial equipment include:

- cocks and valves
- governors
- non-return or back-pressure valves
- low pressure cut-off valves
- relay valves
- solenoid and safety shut off valves
- pressure switches
- flame traps
- fusible links

The list excludes flame protection equipment which is dealt with in the next chapter. Time switches and thermostats may also be used.

A number of the devices listed have domestic applications and were described in Volume 1, Chapter 10.

§74 Cocks and Valves

Manually operated valves used on industrial equipment include:

- plug valves
- butterfly valves
- ball valves
- diaphragm valves
- disc-on-seat valves
- needle valves
- proportioning valves

Plug Valves

These are used in the form of quadrant cocks and thumb cocks. They are used for isolation as well as the control of flow. Heavy duty lubricated plug valves will withstand pressures up to 7 bar ($100 \, lbf/in^2$). The flow through the valve is linear, that is, the flow is directly proportional to the angle of opening.

A cross-section of one type of plug valve is shown in Fig. 38.

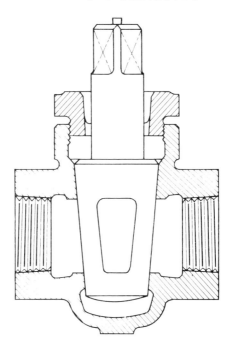

Fig. 38.   Plug valve

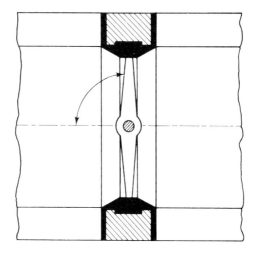

Fig. 39.   Butterly valve; soft seating

Butterfly Valves

These valves are generally used to control low pressure air or gas flows. They have large port openings and so allow high rates of flow. Where complete shut off is required it is essential that a valve with a soft, neoprene-lined seating is used, Fig. 39. This valve can withstand pressure up to 7 bar ($100 \, lbf/in^2$) and may be inserted between flanges in a pipe run.

Gate Valves

This type of valve is usually used for isolating purposes. Because the gate of the valve is lifted completely out of the pipe run when fully open, the valve gives full flow with minimal pressure loss. A parallel slide valve is shown in Fig. 40. Gate valves are slow acting and require several turns of the wheel to close the valve completely.

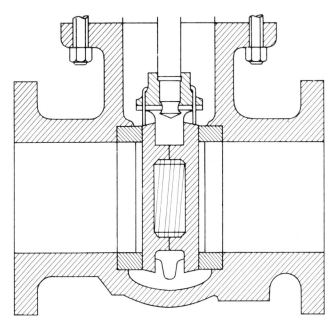

Fig. 40.   Gate valve; parallel slide

Ball Valves

These valves, Fig. 41 may be used either for isolation or for through-put control. With nylon seals the valves can withstand pressures up to about 7 bar ($100 \, lbf/in^2$). Like the plug valve, the ball valve has mechanical over travel. That is, it moves beyond the pneumatically sealed position to a mechanically closed position. When operated automatically it can be used for such controls as the double block and vent system.

Diaphragm Valve

The valve is closed by depressing the flexible diaphragm until it makes contact with the seating. Further compression of the diaphragm

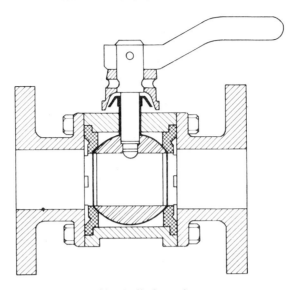

Fig. 41.   Ball plug valve

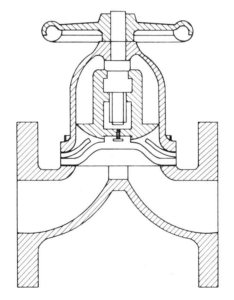

Fig. 42.   Diaphragm valve

ensures complete closure, even when grit or small particles in the fluid lodge under the diaphragm. These valves offer very good control of gas flow with their fine progressive adjustment, Fig. 42.

Because the moving parts of the valve are above the diaphragm,

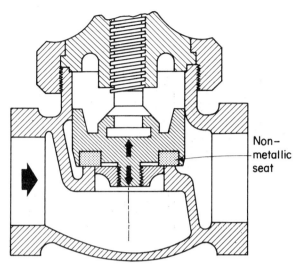

Fig. 43.   Disc-on-seat valve; vertical seat

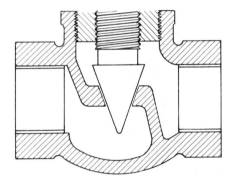

Fig. 44.   Needle valve

they do not come into contact with the fluid. So the valves may be used to control corrosive liquids, provided that these are compatible with the diaphragm and the valve body.

Disc-on-seat Valves

As their name suggests, these valves have a non-metallic disc which closes on to a knife-edged seating. The seat can be either in the vertical or horizontal position. A typical valve is shown in Fig. 43. These valves are used for isolating purposes only and the same disc and seat arrangement is found on automatic valves such as the solenoid, safety shut off and slam shut valves.

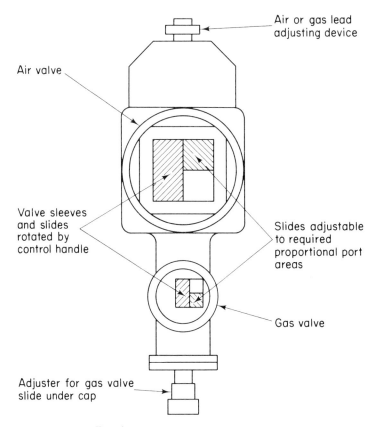

Fig. 45. Adjustable port valve

Needle Valves

Needle valves, Fig. 44, are used where fine adjustments of low gas rates are required. The handwheel withdraws the tapered "needle" out of the similarly tapered seating to increase the flow rate. The valves can be obtained to withstand pressures of up to 28 bar (400 lbf/in^2).

Proportioning Valves

There are two main types of proportioning valves:

- adjustable port valves
- adjustable flow valves

Adjustable port valves. Fig. 45 consist of mechanically linked plug valves on a common spindle or separate valves with connected valve levers. Rotation of the valve spindle controls the throughput. Adjustment of the valve slides controls the proportioning by altering the

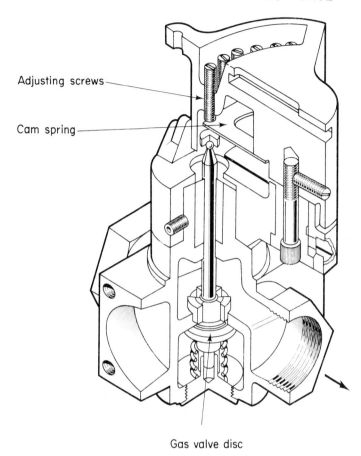

Adjusting screws

Cam spring

Gas valve disc

Fig. 46.   Adjustable flow valve; micro-ratio

height of the rectangular ports to vary the relative port area ratio.

Provision is made for adjusting the position of the spindles at which each valve opens or closes.

Adjustable flow valves are illustrated in Figs. 46 and 47. On both the valves shown the flow rate is varied by the rotation of a cam which alters the amount of port opening. The cam is in the form of a spring which can be adjusted to the required profile by means of a series of screws. In Fig. 46 the cam is rotated to move the valve spindle, whilst in Fig. 47 the cam is stationary and the rocker arm is traversed around it.

The cylindrical piston in Fig. 47 has a rectangular port which is moved up or down by the cam as the piston is rotated, so varying the opening.

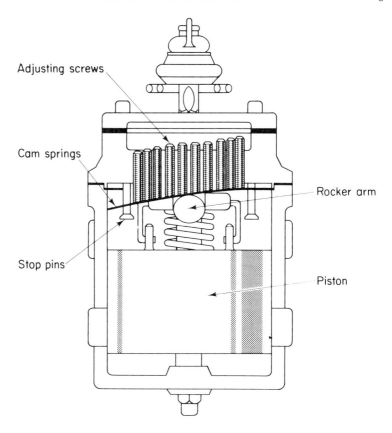

Fig. 47. Adjustable flow valve (Stordy)

§75 Governors

Low Pressure Governors

Gas supplies to industrial or commercial equipment are normally controlled by low pressure governors. These were described in Volume 1, Chapter 6.

The type used is the double diaphragm compensated constant pressure governor. It usually has neoprene main and auxiliary diaphragms, a valve covered with nitrile rubber and spring loading.

The governor is usually the first device in the control train down stream from the isolation valve. A small governor should also be included in the pilot gas supply line.

When volumetric control is required the constant pressure governor is used with back loading from down-stream of a valve or orifice. The governor responds to the differential pressure across the orifice and maintains a constant volume.

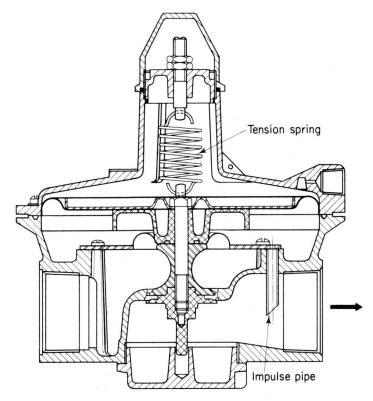

Fig. 48.   Zero governor

Zero Governors

When gas is to be entrained by a stream of air under pressure, its pressure must be reduced to zero gauge pressure, that is, atmospheric pressure. The device used is a zero governor.

This is similar to the constant pressure governor but it has a thin, flexible tension spring in place of the normal stout compression spring, Fig. 48. The tension spring supports the weight of the moving parts, that is, the diaphragms, valve and valve spindle. Small changes in the outlet gas pressure may be obtained by adjusting a nut at the top of the spring support.

Because the top of the main diaphragm is exposed to the atmosphere, via the breather hole, and the spring supports the moving parts, then the downward force acting on the diaphragm is atmospheric pressure. This is balanced by the upward force of the outlet pressure below the diaphragm. So any increase above atmospheric pressure downstream of the valve will cause it to close, whilst any fall in outlet pressure will result in the valve being opened by the pressure of the atmosphere on the main diaphragm.

Zero governors are used in air blast systems and should be installed downstream of a constant pressure governor. They are also used in the pressure divider system of control for nozzle mixing burners. In this case the outlet pressure is not zero but the same pressure as in the impulse line.

§76 Non-return Valves

A non-return valve, sometimes called a back-pressure valve is designed to allow fluids to pass in one direction only. Any reversal of flow closes the valve instantly.

It is a requirement of the Gas Act 1972 that, where air from a fan or compressor, or any other type of gas, is used in conjunction with a gas burner, a non-return valve must be fitted in the main gas supply. This is to prevent air or any other extraneous gases being admitted into the gas service pipe.

A non-return valve must not pass a flow from outlet to inlet exceeding:-

- $0.001 \, m^3/h$ on valves up to 25 mm (1 in)
- $0.003 \, m^3/h$ on valves from 25 mm (1 in) to 50 mm (2 in)
- $0.006 \, m^3/h$ on valves from 50 mm (2 in) to 75 mm (3 in)
- $0.010 \, m^3/h$ on valves from 75 mm (3 in) to 100 mm (4 in)

The valves must be capable of withstanding a reverse pressure as follows:-

- 7 bar ($100 \, lbf/in^2$) on 25 mm (1 in) valves
- 2 bar ($30 \, lbf/in^2$) on valves 25 to 150 mm (1 to 6 in)
- 1 bar ($15 \, lbf/in^2$) on valves above 150 mm (6 in)

Where the air fan is situated above the burner, it is good practice to insert a non-return valve in the air line to prevent gas from entering the air fan casing.

Typical preferred non-return valves are shown in Fig. 49. In the valve at (a), gas entering lifts the two leather diaphragms off their seats on the spring loaded valve head so allowing gas to flow to the outlet. In the event of reverse pressure the diaphragms return to their seating, preventing any return flow past the valve. If the return pressure is increased, the complete metal valve is forced down on to its seating against the spring pressure. This provides an additional seal to withstand the higher back pressure.

The valve must be mounted horizontally.

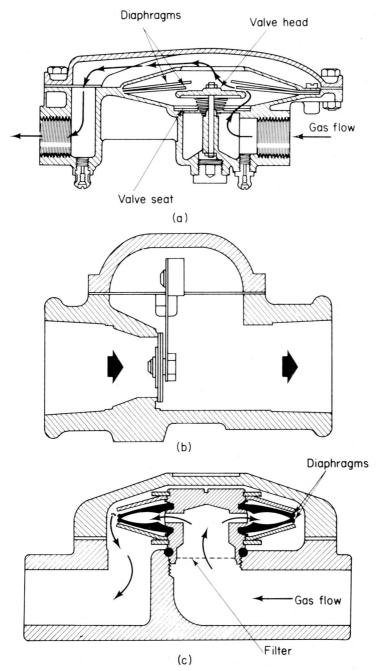

Fig. 49. *Non-return valve (a) double diaphragm (b) roll check valve (Donkin)*
(c) Adaptogas

The valve at (b) is a simple, flap type, disc-on-seat valve with a cast iron body and a knife edge valve support pivot. The valve seat is self-aligning.

The valve must be mounted horizontally, with the dome uppermost.

The valve at (c) has a pair of nitrile rubber diaphragms which touch around the periphery. In forward flow, gas forces the diaphragms apart but under reverse pressure the lips of the diaphragms are forced together so preventing gas flow through the valve.

The valve can be mounted in any position.

If the electronic flame protection system is employed and the safety shut off valve is able to withstand twice the maximum pressure of the extraneous gas or air, applied in the reverse direction, this would meet the requirements of the non-return valve.

Non-return valves should be periodically checked for correct operation and pressure points are incorporated on the inlet and outlet for this purpose. Leather diaphragms and valve seats should be regularly dressed with an appropriate oil to keep them supple.

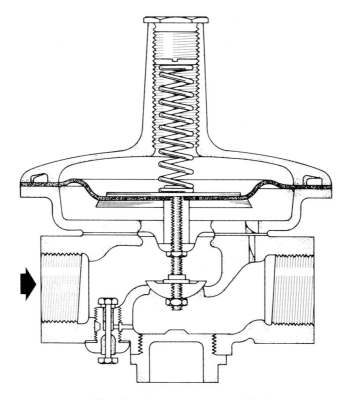

Fig. 50. Low pressure cut-off valve

77   Low Gas Pressure Cut-Off Systems

The object of these systems is to ensure that, when the gas pressure falls to a predetermined value above atmospheric pressure, a valve closes and cannot be reopened until:

- all down-stream burner valves are closed
- the system is restarted manually, for example by depressing a button.

Low Pressure Cut-Off Valves

The valve originally used for this purpose was the low pressure cut-off valve described in Volume 1, Chapter 10. A typical valve is shown in Fig. 50. This has an auxiliary diaphragm so that inlet pressure does not exert any downward force on the valve. It is reset by a weep of gas through the by-pass orifice when the pressure reset plunger is depressed. The size of the orifice is a compromise. It must be small enough to detect a down-stream burner valve left open and large enough so that the operator does not have to keep the plunger depressed for too long. These valves are fairly slow in closing and should therefore only be used on small installations. They are now seldom fitted and are being superseded by low pressure switches in

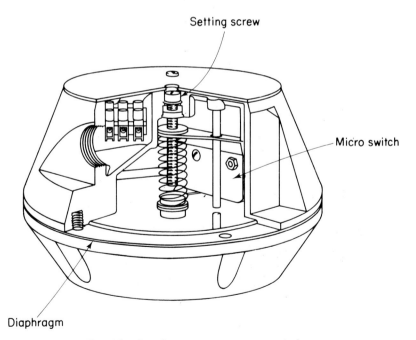

Setting screw

Micro switch

Diaphragm

Fig. 51.   Diaphragm type pressure switch

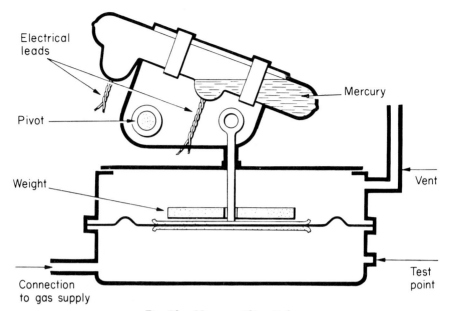

Electrical
leads

Mercury

Pivot

Weight

Vent

Connection
to gas supply

Test
point

Fig. 52.   Mercury tilt switch

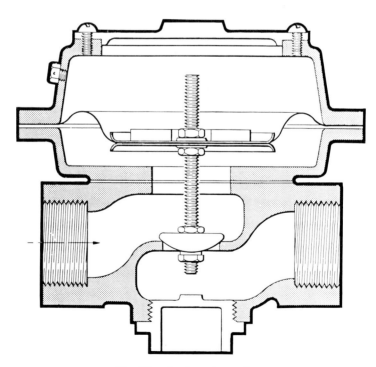

Fig. 53.   Anti-suction valve

conjunction with certificated safety shut off valves or, in some cases, by flame protection systems.

Low Pressure Cut-Off Switches

A diaphragm operated low pressure cut-off switch is illustrated in Fig. 51. This has a micro switch which is held closed by gas pressure acting under the diaphragm against the set tension of a spring. An earlier type uses a mercury tilt switch Fig. 52.

Pressure operated switches may be used for high gas pressure protection by arranging for the switch contacts to remain closed until the set pressure is exceeded. They may also be used for low air pressure protection.

Anti-Suction Valves

Another form of cut-off which has been used to protect gas meters from implosion due to suction from a compressor is shown in Fig. 53. The valve closes when the inlet pressure falls to a set amount above atmospheric pressure. This prevents a suction or reduced pressure reaching the inlet side. The valve is fitted in the gas supply on the inlet of the compressor.

Like other diaphragm operated gas valves, these are being superseded by pressure operated switches controlling safety shut-off valves.

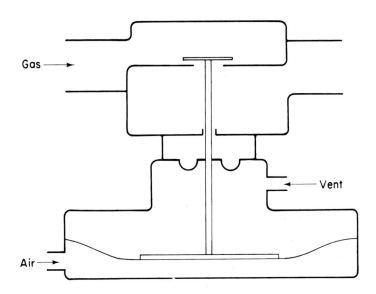

Gas →

Vent

Air →

Fig. 54.   Air flow failure valve; diaphragm operated

§78 Air Flow Failure Devices

The purpose of these devices is to cut off the gas supply if the combustion air supply fan fails. They may be either diaphragm operated gas valves or a pressure switch in the air line controlling an appropriate electrically operated isolation valve.

A diaphragm operated valve is shown in Fig. 54. Failure of air pressure causes the valve to close.

§79 Combined Gas and Air Pressure Failure Systems

It is common practice to combine gas and air failure protection in one system. This may be purely pneumatically operated as in Fig. 55. If either the air or the gas supply pressures fail, the main gas valve will close. When both supplies are restored it requires manual intervention to re-open the valve. The individual burner valves must all be closed and then the button on the manual reset valve must be depressed to build up pressure under the auxiliary valve diaphragm.

One of the shortcomings of the pneumatic system is that it is relatively slow in operation. This drawback is overcome by the electric system in Fig. 56.

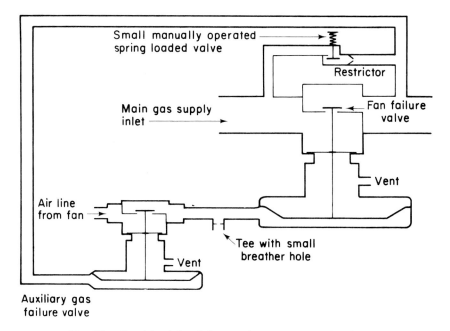

Fig. 55. Combined fan failure and pressure cut-off valve

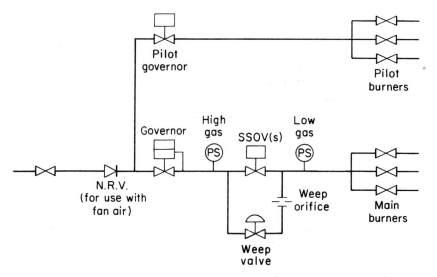

Fig. 56. Low pressure cut-off system

Pressure operated switches in the gas and air lines are wired in series with the safety shut off valve in the gas line. If any switch opens, the safety shut off valve will close. A switch will open in the event of:

• low air pressure or
• low gas pressure or
• high gas pressure

The gas supply can only be restored when the air fan is operated and by closing all gas burner valves and depressing the weep valve to actuate the low gas pressure switch.

On high temperature plant operating at temperatures of 750°C (1380°F) or above, a low gas pressure protection system should be used if a flame safeguard is not fitted. This is to comply with the requirements of the Code of Practice for the use of Gas in High Temperature Plant (Chapter 5).

§80   Relay Control Valves

The operation of relay valves was described in Volume 1 and a typical industrial type is shown in Fig. 57. This may be used to control either air or gas flow. Various shut off valves may be fitted in the weep line, the most common on smaller installations are indirect acting thermostats. Solenoid valves are also used, frequently in conjunction with a thermostat switch or a time clock.

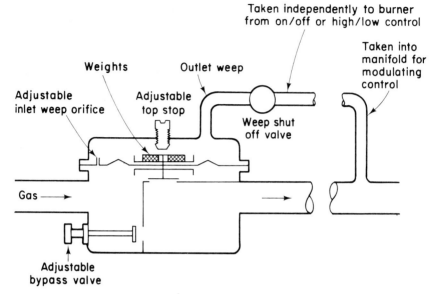

Fig. 57. Relay control valve

There are three methods of operation:

· on/off
· high/low
· modulating

On/off operation takes place with the adjustable by-pass valve closed. The weep gas is burned independently adjacent to the pilot burner. The by-pass valve may take the form shown or be an adjustable stop below the valve to limit its closing.

High/low operation uses the by-pass valve to adjust the low-fire rate. If an adjustable top stop is fitted this is used to adjust the high-fire rate. The weep gas is again burned independently adjacent to the pilot burner.

Modulating control is provided with the by-pass valve closed and the weep gas returned to the gas manifold down-stream.

Relay valves provide a very cheap method of proportional control on small burner systems.

§81 **Solenoid Valves**

Solenoid valves were introduced in Volume 1, Chapter 8. They are commonly used for electrically operated flow control and in certain circumstances as safety shut off valves. Solenoid valves open or close immediately the current is switched on or off.

When used as safety shut off valves, fast opening direct acting solenoid valves should satisfy the requirements of the "Standards for Automatic Burners, Forced and Induced Draught". For the valve to be certificated by M.R.S. it must pass a life test of one million operations for valves up to and including 40 mm ($1\frac{1}{2}$ in) connection size and 250 000 operations for valves with over 40 mm ($1\frac{1}{2}$ in) connections. Valves must close in 1 second or less.

The compression springs on large solenoid valves can exert only a low closing force. This makes them unsatisfactory for use where positive closing must be ensured. Also, under certain conditions their fast opening may be undesirable. A sudden surge of gas may extinguish pilot burners.

Because of these characteristics, solenoid valves are more frequently used to control small burner systems or pilot lines on larger burners.

§82   Safety Shut Off Valves

One of the requirements of the "Standards for Automatic Burners", is that these valves should be relatively slow opening with a time range for main burner valves of 2 to 5 seconds. There are two types of slow opening valves:

- electro hydraulic
- electro mechanical

Both these systems enable the valve to open slowly against the closing force of a strong valve spring to ensure positive shut off. The valves must close in one second or less and, to be certificated and appear in the six monthly issue of the Certificated Equipment List, they must satisfy the life test previously described.

Electro Hydraulic Valve

This valve is described in Volume 1. An electric motor drives a pump to raise the pressure of oil above a diaphragm which moves to open the gas valve. The valve is held shut by a powerful compression spring. When current is switched off an electromagnetic relief valve opens allowing the oil to flow quickly back into the reservoir and the valve closes.

Electro Mechanical Valves

One type has an electric motor driving a rack and pinion through a magnetic clutch. The rack forces the valve open against the tension of a strong spring. When de-energised, the clutch is disengaged and the valve is closed by the spring.

In another type, Fig. 58, the electric motor turns a cam. The cam

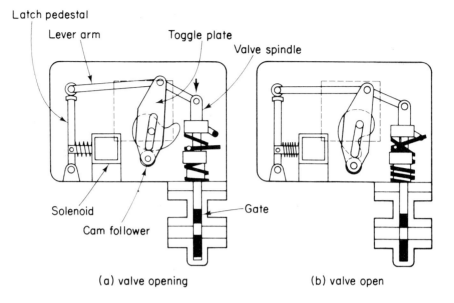

(a) valve opening   (b) valve open

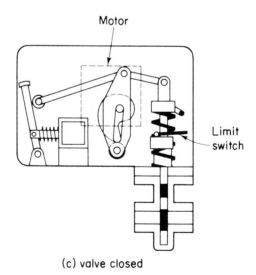

(c) valve closed

Fig. 58. Safety shut off valve

follower is connected, by a toggle plate and a lever arm, to the valve spindle. A solenoid is incorporated which when energised, provides a fulcrum for the lever arm on top of the latch pedestal. When the valve is energised, the latch pedestal is vertical and the lever arm opens the gate valve against the tension of the compression spring. A limit switch controls the amount of valve travel.

When current is switched off, the solenoid is de-energised, the latch pedestal moves away, allowing the lever arm to drop and the spring to close the gas valve.

§83   Flame Traps

Where premixed air and gas is used in gas fired equipment a flame trap should be fitted to prevent a flame from passing back through the pipework to the mixer. A satisfactory flame trap should do two things:

- stop the flame passing through the pipe system
- cut off the fuel supply if a flame reaches the flame trap

A typical flame trap element, Fig. 59, is built up in a similar way to the ribbon burner from alternate strips of flat and corrugated steel ribbon, wound in a spiral. It has the effect of breaking up the flame over the large surface area and quenching it in the very small passages between the corrugations.

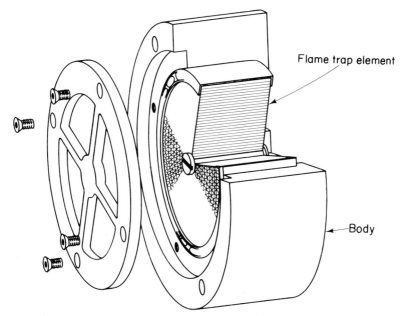

Fig. 59.   Flame trap element

Flame traps should be situated as near as possible to the gas burner. This is so that the flame does not have a long pipe run in which it might accelerate to such a speed as to form a detonation wave and make the trap useless.

If a flame was allowed to continue burning against the element it might heat the trap until the mixture ignited on the upstream side. To avoid this thermocouples can be placed on the burner side which will sense the presence of a flame and close a gas valve via temperature detector contacts.

Flame traps are used in other situations including:

- pressure relief systems and governor vents
- purge outlet pipes

§84 Fusible Links

These are a form of limit control which may be used to:

- prevent accidental overheating
- protect premix systems against lighting back
- control a process by shutting off the gas when a given temperature is reached

The devices may be either mechanical or electrical. Mechanical devices may employ the link as part of a chain supporting a weight which, when released, closes a gas valve. Another type holds a gas valve open against the tension of a spring.

Electrical control may be effected by fitting a fusible link as a thermocouple interrupter. When the link melts the circuit is broken.

Because softening occurs before the melting point is reached the links are liable to cause premature shut down and have largely been superseded by other more accurate temperature controls.

Fusible plugs are however required to be fitted to some steam boilers by the Factories Act. This stipulates that every steam boiler, unless externally fired, shall be fitted with a suitable fusible plug or an efficient low water alarm device. The Certificate of Exemption No 18 waives this requirement for small multi-tube boilers not more than 3 ft (900 mm) diameter, fired by gas or oil.

Fusible plugs act as an alarm by releasing steam and water when local overheating occurs, but they do not shut off the gas or oil supply. BS 759 consequently recommends that fusible plugs should not be fitted to oil or gas fired boilers which should instead have low water alarm devices.

When changing over a solid fuel boiler to gas firing any fusible plug should be removed and a low water alarm fitted (Chapter 11).

§85 Servicing

To maintain gas fired plant in operation at its optimum performance and to ensure that all the safety features in its control system are always effective, planned, regular servicing is essential. A routine programme should be designed so that equipment may be serviced when it is least likely to be needed for production.

Where equipment has been in use for several years it should be updated to incorporate the appropriate requirements of relevant Standards and Codes of Practice during periodic servicing.

The items of plant to be serviced include:-

- burners
- fans and boosters
- manual valves
- safety shut off valves
- non-return valves
- governors
- throughput and mixture control
- flue gas controls
- flame protection equipment
- electrical equipment

Burners

For burners where the air/gas ratio has been correctly maintained, very little servicing is normally required.

On natural draught systems the injectors and burner parts should be checked to ensure that they are not clogged with dust which would reduce the primary aeration. Auxiliary flame parts on bar or ring burners should be free from deposits or scale which could render them ineffective.

On forced draught systems, make sure that burner quarls are in good condition and that no loose refractory material obstructs the entrance to the combustion chamber. Check all flames visually for stability and correct flame profile.

Fans and Boosters

These items of equipment should be checked in accordance with manufacturer's instructions.

Lubrication should only be carried out if specified and with the correct oil or grease. Check for excessive noise or vibration and ensure that the mountings are secure and in good condition. Check that the air fan inlets are not obstructed by dust or dirt and clean out if necessary. Ensure that all glands and seals on gas boosters are sound and that there is no smell of gas in the area. Ensure that warning notices are correctly displayed.

Manual Valves

All manual isolating valves should be pressure tested annually to ensure that they are leak tight. This is done with a gauge fitted between closed valves. If there is a perceptible rise in pressure the faulty valve should be exchanged or repaired.

The servicing requirements of the various types of valve are as follows:

Plug Valves

The low pressure gas and air quadrant cocks and thumb cocks should have the plugs lightly greased to ease their operation. If the valves leak the plug should be lapped in with a fine paste, greased and re-tested.

Lubricated taper plug valves for use at higher pressures should be filled with a slug of grease periodically, depending on the frequency of their use.

Ball Valves

The majority of these valves have nylon ring seals which take up any wear. The servicing required is minimal.

Gate Valves

These valves have a tendency to jam in the open position if not frequently used. It is therefore important that they should be checked at six month intervals to ensure that they open and close freely. The gland seal should be tested with leak detection fluid to ensure that there are no leaks around the valve spindle.

Diaphragm Valves

These should be checked annually for correct operation. The valves should be dismantled about every three years so that the diaphragm may be examined for wear and replaced if necessary.

Safety Shut Off Valves

All certificated safety shut off valves and normally open vent valves must be pressure tested at least annually or more frequently, depending on how much the plant is used. Single valves should be tested against a downstream manual valve. The space between the valves is vented, then any perceptible rise in gauge pressure between the valves over two minutes indicates the need for further investigation. On some of the solenoid valves, removal of the bottom cover reveals the valve and seating. Any foreign matter should be carefully removed, the valve seat cleaned with a non-fluffy material, the cover replaced and the valve re-tested. On the electro-hydraulic valves access to the valve and seating is gained by completely removing the actuator, leaving the valve body in the line.

If the actuator fails to open the valve it should generally be returned to the manufacturers for repair. So spare actuators or valves should always be available to maintain the continuity of the production process.

With two safety shut off valves in series, as for example on automatic packaged burners, the space between the valves should be vented through the test point and then checked for pressure by a gauge connected to the test point. If the pressure rises above atmospheric, the upstream valve is faulty.

If the test is satisfactory, the space between the valves is then pressurised either by an air pump or with line pressure via a temporary flexible tube connection. When the pressure is established there should be no rise in pressure between the second safety shut off valve and the final manual isolating valve. If this pressure does rise, the second safety shut off valve is leaking.

On large burner systems a normally open vent valve is fitted between the two safety shut off block valves. To check the vent valve, first check the two block valves then energise the vent valve by connecting it to an external electrical supply and pressurise the space between the three valves. If pressure is lost over a period of two minutes, the valve must be repaired or re-placed. Finally reconnect the vent valve to its electrical supply and check that it operates in the correct sequence.

Non-Return Valves

The most widely used valve is the double diaphragm type shown in Fig. 49(a). Because the diaphragms and the valve seat are made of leather they should be dressed with a light oil of a type recommended by the manufacturer to keep them from drying out in the presence of natural gas. This is carried out by removing the top cover to expose the diaphragms and by removing the large plug beneath the valve to dress the valve seat.

The disc-on-seat flap valve, Fig. 49(b), should have the seat cleaned of any foreign material whilst the rubber diaphragm type, Fig. 49(c), should require very little servicing.

Annually the effectiveness of all non-return valves should be checked. If possible remove the valve from the line and carry out a reverse flow check against a closed valve using a bubble test meter. If there are more than one or two bubbles per minute the valve must be replaced.

Air and Flue Dampers

Some sectional boilers with natural draught burners are fitted with safety shut off valves with a lever which operates the air inlet damper. When the valve is closed the air inlet louvres should be in the closed position. Check that when the valve is open the louvres are in the fully open position. Reset and tighten the air linkage mechanism as necessary.

Check the operation of any flue dampers. Ensure that the damper flap is in good condition and not corroded away inside the flue. Check that the damper can be correctly adjusted.

Governors

Governors normally require minimal servicing. The large governors fitted upstream of the primary meter are the province of specialist staff. Individual appliance governors should be examined to ensure that diaphragms are sound and in good condition. Valves and seatings should be cleaned and any rust or dust deposits removed. Check that impulse pipes are clear, sound and secure. The outlet pressure should be reasonably constant between low and high fire rates.

Throughput and Mixture Control Systems

Many low temperature, low thermal input appliances such as vats and tanks, air heaters and sectional boilers are controlled by relay valves. These may give on-off, high-low or modulating control. They should be checked for operation by adjusting the thermostat and noting the response of the burner. Ensure that weeps and orifices are clear, valves clean and diaphragms sound. Check that top and bottom stops for high and low fire rates are correctly set and tightened.

The mixture control on many industrial furnaces is the premix air blast system. Check that at low fire rate, the zero governor outlet pressure is atmospheric. Adjust the tension spring or service the governor as necessary. Check at high fire rate and adjust the injector obturator screw as necessary to give the required flame. Retighten all locking nuts after adjustments have been made.

Nozzle mix burner control is commonly linked by butterfly valves. Check the linkage between the valves and examine the flame at low and high fire. Adjustment of the top and bottom stops of the valves should be made to give the correct aeration. If a pressure loading system is employed, adjust the zero governor tension spring and the trimming valves to obtain correct aeration at low and high fire rates.

Larger burners with full modulating control use either adjustable port valves or adjustable cam profile valves. In the first system check the port settings and any linkages to ensure that a stoichiometric air/gas mixture is maintained over the turn down range. In the second system, check the tightness of the cam profile adjusting screws and any wear in the Bowden cable or other linkage between the valves.

Unless the air and gas supplies are metered, the normal check on throughput control equipment is by recognition of the sight and sound of a correctly aerated flame. This ability may be developed as the result of experience.

A more objective test of the completeness of combustion and the efficiency of the appliance is by carrying out a flue gas analysis.

Whenever possible this should be taken at the various burner settings. With natural draught installations the products of combustion measured at points across the flue should contain at least 4% CO_2 when measured by a Draeger or Fyrite analyser. On forced draught systems there should be a minimum of 9% CO_2 at all settings with a maximum of 6% O_2 which may be checked with either a Fyrite or a Servomex paramagnetic instrument. In all cases there should be no trace or only minimal traces of CO when checked by a Draeger tube.

On boilers the maximum permissible CO concentration is 100 p.p.m. for non-shell boilers and 200 p.p.m. for packaged shell boilers with a water cooled probe at the back end of the boiler.

Flame Protection Equipment

This equipment and its servicing is dealt with in the next chapter.

Electrical Control Equipment

All electrical equipment should be checked annually to ensure that all switches and interlocks function correctly. Gas and air flow and pressure switches are particularly vulnerable. Pressure switch diaphragms may rupture, pressure settings vibrate out of adjustment and micro switch contacts weld together. Check that normally closed contacts are closed when the system is at rest. Artificially reduce the air or gas pressure being monitored by the switch and check with a gauge that, when the pressure falls below the setting level, the appropriate interlock opens and shuts the system down.

Position proving switches on valves, doors and dampers should also be checked to ensure that, if the device is not in the correct position, the switch will be either open or closed as required by the circuit to provide protection. Similarly, on boilers and water heaters, high, low or extra low water level switches, steam overpressure or temperature overheat limit switches should be checked. A multimeter should be used to establish that contacts are in their correct position and function satisfactorily.

Time clocks should be manually rotated to ensure that the process or heating system switches on and off at the appropriate times. Temperature and over-heat settings should be checked to ensure that the systems operate at the desired temperatures. No system should be run at a temperature higher than that absolutely necessary for the purpose, otherwise fuel will be wasted.

Semi-automatic and fully automatic programming control units should be checked annually by running the system through its sequence and noting the time intervals for each stage of the cycle. Simulate loss of flame and loss of air and gas pressures. The unit must go to lock out and shut down, closing the safety shut off valves immediately.

If a system fails to start up it is usually an interlock which is at fault, rather than the control unit itself. Check all interlocks first and when these have been proved to be satisfactory, isolate the equipment electrically and remove the plug-in control unit from the fixed base. Thoroughly clean the spring loaded connection contacts between the base and the control unit with fine emery paper if necessary. On most units the chassis pins are self cleaning and removing and replacing the unit cleans the contacts. If, when replaced, the unit still fails to operate a new plug-in control unit should be fitted to the base. No attempt should be made to repair the faulty unit which should be returned to the manufacturer for servicing. To ensure continuity of the heating process, a spare control unit should be kept by the customer.

Any flexible metal protection sheathing which is fitted to field interlocks such as position or pressure proving switches must be made secure. The earth continuity bonding of all controls, interlocks, motors and switchgear should be checked annually, preferably with a Megger.

§86 General Notes

Since the introduction of natural gas, many processes have been changed over to its use. New burner systems are continuously being developed and new Standards and Codes of Practice devised to ensure that plant is operated as efficiently and safely as possible. A list of current standards and codes associated with industrial gas burners and controls is given in the Appendix. These are continuously under review and it is likely that a number of the recommendations quoted in this chapter may be out dated by the time it comes to publication.

§87   **List of Relevant Standards and Codes of Practice Relating to Industrial Gas Burners and Controls**

Standard for Automatic Gas Burners (Forced and Mechanically Induced Draught) Publication IM/8 July 1977.

Code of Practice for Large Gas and Dual Fuel Burners. Report IM/7 May 1976.

Code of Practice for the Use of Gas in High Temperature Plant. Publication IM/12 February 1980.

A Guide to the Gas Safety Regulations Report IM/4 April 1975.

Soundness Testing Procedures for Industrial and Commercial Gas Installations Report IM/5 July 1979.

Purging Procedures for Non-Domestic Gas Installations Report IM/2 February 1975.

Flues for Commercial and Industrial Gas Fired Boiler and Air Heaters, IM/11, May 1979.

Combustion and Ventilation Air Guidance Notes for Boiler Installations in Excess of 2,000,000 Btu/h (586 kW) Output 1975.

Code of Practice for the Use of Gas In Atmosphere Gas Generators and Associated Plant Parts 1, 2, 3 IM/0 October 1977.

Technical Notes on Changeover to Gas of Central Heating and Hot Water Boilers for Non-Domestic Applications April 1978. Publication IM/10.

Recommendations for the Installation of Low Pressure Cut-off Switches to Conform with the Gas Act Report IM/6 September 1974.

Manual Valve Descriptions and Selection Recommendations for Industrial and Commercial Applications (Draft) Report IM/105 June 1976.

Evaporating and Other Ovens, Health and Safety at Work Series No. 46 Department of Employment, HMSO.

Draft Code of Practice for the Use of Gas in Low Temperature Plant, Publication IM/103, 1979.

Non-Return Valves for Oxy-Gas Glass Working Burners, Publication IM/1, 1972.

Flame Protection Systems

Chapter 5 is based on an original draft by Mr. W. A. Pidcock

§88   Introduction

The purpose of flame protection systems is to safeguard gas fired equipment from hazard during any phase of its operation. These phases are:

- start up
- normal run
- shut down

The most critical phase is that of starting up, so, as well as giving protection from flame failure at any time, the system must be linked to the ignition procedure.

Commercial appliances use a variety of flame protection devices. These range from domestic models on the small appliances to industrial models on large equipment. Flame protection systems should possess a number of indispensable characteristics and should generally not be used on industrial equipment unless they comply with the following requirements.

Flame protection systems should:

- ensure that the appropriate lighting up procedure is correctly applied before the burner will light
- either   prevent any gas being supplied to the main burner until the pilot flame is established
- or   prevent the full gas rate being supplied to the main burner until a flame at a low rate has been established and sensed for a trial period
- stop all gas being supplied to the burners after flame failure and require manual resetting; unprotected pilots should not be used with devices for industrial applications
- be adjusted to sense only that part of the flame which will ignite the main burner and not respond to any other flame or a flame simulating condition.
- in the case of electronic devices, be provided with a "safe start

174

check" to prevent energising the gas valves and the ignition if a 'flame on' condition exists before ignition.

- thermoelectric types should ensure that the main gas valve is manually isolated until the pilot is established.
- be mechanically and electrically sound and readily serviced.
- when correctly installed, be free of any tendency to fail to danger.

In addition to these essential characteristics there is a number of other features which are desirable.

Flame protection systems should, if possible:

- be protected against interference by unauthorised persons
- operate satisfactorily under all changing conditions
  of—throughput
     —flue draught
     —mixture ratio
     —gas characteristics
- operate satisfactorily within the ambient temperature range to be expected on the appliance
- where powered by mains electricity, operate satisfactorily within supply voltage variations of +10% and −15% of the nominal rating
- be unaffected in its operation by foreign matter
- tolerate reasonable vibration and shock
- where it includes a gas carrying component, pass the required volume of gas within the permitted pressure loss

§89   Ignition Devices

Because flame protection systems are linked to the appropriate lighting procedures, it is useful to review the methods by which industrial and commercial appliances may be ignited.

These are:

- manual ignition—by match or taper
                          —by lighting torch
- pilot flame
- electric ignition—hot wire filament
                          —piezo electric spark
                          —mains transformer spark
                          —electronic pulse spark

Ignition and ignition devices were dealt with in Volume 1, Chapters 2 and 10 respectively.

Manual Ignition

Hand held matches or tapers are used on domestic appliances and may be used on some of the smaller commercial appliances. For industrial appliances the lighting torch is recommended. This is typically a length of 6 mm ($\frac{1}{8}$ in) m.s. pipe connected to a piece of flexible tubing. The length of the torch must enable the operator to ignite the burner while within easy reach of the burner control valve. The gas rate of the torch must not be more than 7.3 kW (25 000 Btu/h) or 3% of the main burner gas rate, whichever is the greater.

Pilot Flames

Pilot flames may be:

- continuous burning
- intermittent
- interrupted

A continuously burning pilot remains alight while the appliance is in operation, whether the main burner is on or not. Some pilots are increased in size when required to light the main burner. They may be reduced after ignition or when the main flame is shut off.

An intermittent pilot is lit prior to ignition of the main flame and is shut off simultaneously with it.

An interrupted pilot is lit prior to ignition of the main flame and is shut off when the main flame is established.

All pilot flames should:

- be positioned to provide immediate and smooth ignition of the main flame at any normal gas rates
- be situated to avoid interference from draughts or combustion products
- be provided with access for ignition and facility for viewing.

Where flame protection is provided, the pilot connection should be taken from the main supply downstream of the main inlet valve.

The pilot burner must be securely mounted to avoid movement affecting its operation. The pilot supply should be rigidly fixed to prevent vibration occurring which could extinguish the flame.

A start-gas flame may be established either at the main burner or at a separate pilot burner. The rate for natural gas shall not exceed 25% of the stoichiometric gas rate of the burner to be fired. The energy release shall not be more than 53 kJ/m³ of combustion chamber volume for every 100 mbar pressure rise the combustion chamber can withstand.

Electric Ignition

Piezo-electric igniters are used on some commercial appliances but are generally not used industrially.

Hot wire filaments in the form of glow-coils are used mainly in heating appliances, in conjunction with mains transformers. Their use is generally restricted to standard appliances with natural draught burners.

Mains transformer ignition may be produced by a step-up transformer. This may either have the electrodes connected to each end of the secondary winding and have a central earth tapping or, more commonly, have one end of the winding connected directly to earth and the other connected to the spark electrode.

It is not possible to obtain sparks simultaneously at two gaps, using the centre earth tapped transformer with each electrode sparking to earth. Sparks will be obtained at the shorter gap or at random at both gaps.

Ignition transformers should be mounted near to the burners but protected from heat or damage. The high tension lead should be insulated from contact with other components or people and not run near any flame detector leads. Electrical pick up can cause signals which simulate a flame.

Spark gaps may be from 2.38 to 9.52 mm ($\frac{3}{32}$ to $\frac{3}{8}$).
Power input is normally 100 to 200 W.

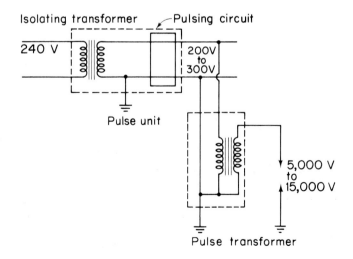

Fig. 1   Electronic pulse ignition circuit

Electronic pulse igniters produce sparks by injecting a high speed pulse of electrical energy into a transformer. This may often be an auto-transformer of the car ignition coil type. The input pulse may be obtained from a solid state switching circuit controlled by a thyristor which triggers repeatedly to discharge a capacitor into the pulse transformer. A single pulse transformer is shown in Fig. 1. but, with the appropriate circuitry, up to 100 transformers can be operated from one pulse generator.

Power input for a single spark unit is about 40 W and for multiple units about 2 to 5 W per spark.

FLAME PROTECTION DEVICES

§90 Type

Flame protection devices may be divided into four main categories, as follows:

- thermal expansion
- thermo electric
- flame ionisation
- photosensitive

Of these, the devices in the first category are not used for industrial equipment and generally only the liquid expansion and vapour pressure types are used commercially. The other three categories are used where appropriate for commercial or industrial applications.

§91 Thermoelectric Devices

The simple direct acting gas valve type was described in detail in Volume 1, Chapter 10. These devices are generally more reliable than the thermal expansion types but are much slower to react than the electronic devices. They are relatively cheap and those which operate the gas valve do not require an electrical supply. They are used extensively for low temperature applications and commercial appliances.

When starting up, the push button should not need to be depressed for more than 30 to 40 seconds. When the flame fails, all gas to the pilot and the main burner should be shut off within about 45 seconds. In operation this is usually within 20 seconds.

There are two types of thermo electric devices:

1. the gas valve is attached directly to the armature of the electric magnet. Thermo electric valves are fairly slow in operation and, in the 45 seconds allowed between loss of flame and closure of the valve, gas is passing into the combustion chamber. These devices should not be placed adjacent to surfaces with high thermal inertia as this would keep the thermocouple warm for a longer period. They should only be used on low throughput, non-automatic natural draught appliances.

2. the magnet is used to operate an electric switch used to control a solenoid or safety shut off valve. This type requires an electrical supply. The second type is less commonly used, electronic devices are usually preferred.

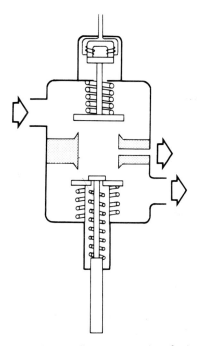

Fig. 2   Thermo-electric flame protection device; gas valve type

Devices and Systems

The direct gas valve type was described in Volume 1 and is shown in Fig. 2. Its application to protect a single bar burner is given in Fig. 3.

A similar device is illustrated in Fig. 4. This has a separate pilot gas supply controlled by a separate pilot valve independently housed and

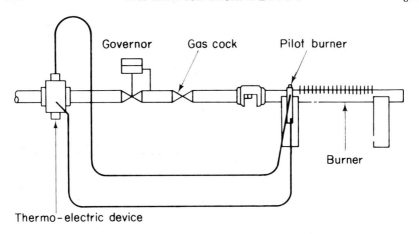

Fig. 3 *Direct protection to a bar burner*

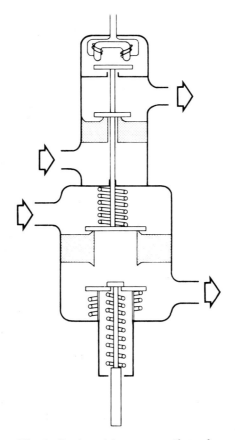

Fig. 4 *Device with separate pilot valve*

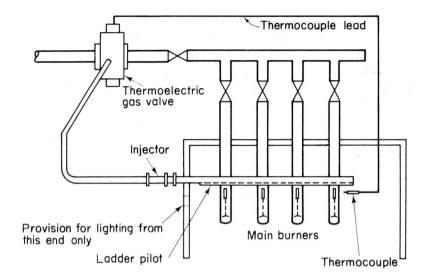

Fig. 5   Ladder pilot on multiple burners

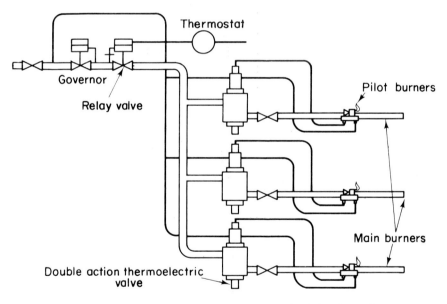

Fig. 6   Protection of multiple burners by separately fed pilot valve on each burner

mounted on the same spindle as the main gas valve and the armature.
  Multiple burner systems have always presented problems and two methods of providing flame protection are shown in Figs. 5 and 6.
  The first is a ladder pilot system using one direct acting device to

supply the main burners and the ladder pilot. This system has the disadvantage that it can fail to danger if the ladder burner becomes blocked over part of its length.

The second system uses individual devices of the separately fed pilot type for each main burner and is a more positive system.

Installation

When installing thermo electric devices the following points should be noted:

- the device containing the gas valve should be located downstream of the main gas valve and upstream of any controls using a weep line
- the pilot supply should be taken from a point upstream of the main governor
- the device should be located so that an operator can depress the start button and easily light and observe the flame
- it should be fitted so that the gas valve moves vertically downwards when closing, assisted by gravity
- the pilot and thermocouple should be located in accordance with the manufacturers instructions and shielded from draughts
- the thermocouple should be protected from radiation from sources other than the flame, the bracket should be attached so as to conduct heat away from the thermocouple
- care must be taken to ensure that the correct temperature difference is maintained between the hot and cold junctions, the pilot and thermocouple assembly should not reach a temperature of more than 300°C (570°F) to avoid overheating the cold junction
- the gas valve and electromagnet assembly should be rigidly mounted to avoid being affected by shock or vibration

Commissioning and Servicing

After installation and again when servicing the system, it must be checked to ensure that it will operate satisfactorily at all normal conditions of draught, throughput and mixture ratio change. A pilot turn down test should be carried out to ensure that, when the pilot is turned down to a point at which it will just hold up the main gas valve, it will still light the main burner satisfactorily.

Procedure for the test is as follows:

- turn off the burner valve
- turn down the pilot flame to a point at which it will just hold up the gas valve; each setting should be held for at least 3 minutes, to reach thermal equilibrium

- check visually that the igniting flame is long enough to light the main burner
- turn on the burner valve and check that ignition is smooth and satisfactory
- repeat the test at the extremes of draught throughput and mixture ratio which might normally occur

A shut down test should be carried out to ensure that the device is left operating satisfactorily. It should also be carried out periodically and might, with advantage, be adopted as part of the normal closing down routine. The procedure is as follows:

- run the appliance up to normal temperature
- turn off the main isolation valve
- note the time taken for the gas valve to shut off by listening or feeling; this should not exceed 45 seconds
- record the time taken when commissioning and check on subsequent tests that the time taken remains substantially the same
- complete the shut down routine by closing burner valves

Instead of listening for the 'click' when the valve closes it is possible to obtain a visual indication by turning the isolating valve down until only small beads of flame remain on the burner. At this point the pilot will have shortened so that it does not materially heat the thermocouple. Note the time taken for the beads of flame to start to go out, indicating that the valve has closed. Complete the shut down by closing the burner valves and the isolation valve.

Further information on servicing thermo electric devices will be found in Volume 1, Chapter 10. Testing of thermocouples is in Volume 2, Chapter 13.

92 Flame Conduction and Rectification Devices

These devices require a 240 V a.c. mains electrical supply. They employ a flame electrode as a sensor and have an electronic amplifier to make it possible for the small current involved to operate a relay. Their advantages over thermal expansion or thermo electric devices are:

- they detect only a flame and are not actuated by heat
- they have a quick response to the presence of a flame or to flame failure
- they can be positioned to prove the presence of a pilot flame at a point where the main flame must be ignited
- the electrode has a longer working life

Both systems rely on the ability of a flame to conduct electric current. During combustion, large numbers of free electrons and ions are present in the flame, so the flame acts as an electrolyte in which a current can flow. The ions and electrons are attracted to suitably charged electrodes and currents of about 10 to 12 μA are conducted.

Flame Conduction

This method used a steady d.c. potential applied between the flame electrode and the earthed burner. The small current conducted was then amplified directly to operate a relay.

The method had a major disadvantage. A build up of dust or condensation on the insulator of the flame electrode can provide a resistive path to earth which results in a current flow indistinguishable from that produced by the flame. Refinements were devised to overcome the problem but it is electronically simpler to use the ability of a flame to rectify a current.

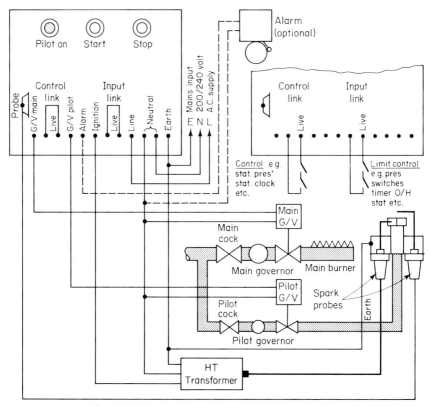

Fig. 7   Flame rectification wiring diagram

Flame Rectification

This method is not liable to flame simulation and has largely superseded the flame conduction device.

It was discovered that, if the earthed portion of the burner nozzle in contact with the flame is considerably larger in area than the flame electrode, then more positive ions will strike the earthed burner nozzle when it is negatively charged than will strike the flame electrode when it becomes positively charged. So more current flows when the burner is negative than when it is positive and a partially rectified current results.

A suitable circuit completes the rectification and the d.c. current is amplified to operate a power relay. (See Fig. 7.)

Any dust build up does not affect the rectification and cannot simulate a flame.

The rectification depends on an adequate area of burner nozzle in contact with the flame, not less than 4 times the area of the flame electrode in the flame. Where this is insufficient the area may be increased by fitting extension rods, a wire spiral or fins to the burner nozzle, Fig. 8. The need for an increase in area is shown by a low milliamp reading when checking the flame signal as instructed by the manufacturer.

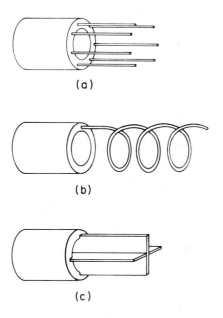

(a)

(b)

(c)

Fig. 8 Extensions to earthed electrode: (a) extension rods; (b) wire spiral; (c) fin assembly

Installation

The successful application of flame conduction and rectification devices is dependent on:

- the correct location of the flame electrode within a suitable flame (see Fig. 9)
- the provision of an adequate earth surface within the flame

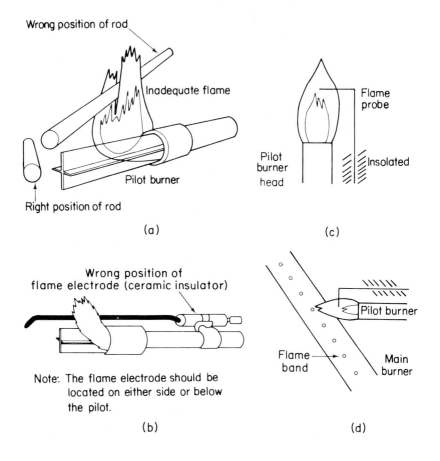

Fig. 9   *Positioning of flame rectification detectors*

The flame electrode should:

- be situated so that, to be detected, the pilot flame must always be of sufficient size to ignite the main burner under all normal working conditions
- be located so that sparks from the igniter cannot arc to the electrode and damage the elctronic components

- be made of a material suitable for use at the operating temperature required
- preferably be a short, straight rod, mounted above and to one side of the main burner to avoid problems from the electrode drooping or sagging
- have the insulator located so that it will not reach a temperature above 350°C (660°F)
- where the lead is subject to high temperatures, employ suitable materials as insulators (glass fibre, P.T.F.E. and silicone rubber may be used up to 200°C (390°F)

The installation should be in accordance with current standards and the manufacturer's instructions.

Commissioning

The installation should be checked to ensure that it meets the recommended requirements. Tests should be made to ensure that the electrode is correctly positioned and that the system operates satisfactorily. When closing down the appliance it is usual to check the operation of the controls by carrying out a shut down test as follows:

- run the appliance up to temperature
- turn off the main isolating valve
- check that the safety shut off valve closes within 1 second or that the flame relay opens with 1 second
- complete the shut down procedure

Servicing

Regular servicing is essential to avoid breakdown and ensure safety. Manufacturers' recommendations should be followed. Generally, the following operations should be carried out.

Weekly

Carry out a shut down test. Also, with the burner firing normally disconnect the electrode lead and check that shut down occurs within 2 seconds.

Monthly

Ensure that the flame electrode is properly located with respect to the main burner and the pilot burner.

Six-Monthly

Clean the flame electrode insulator and check that the lead is in good condition. Check that the burner nozzle earth electrode material, if fitted, is in good condition. Replace any defective parts. Check that flame relay contacts are clean and operating satisfactorily.

Annually

Renew flame and earth electrodes and electrode leads.

§93   Photo-Sensitive Devices

These devices require a 240 V a.c. main electrical supply. They employ a photo-electric flame sensor, sometimes called a "head" or "scanner" in conjunction with an electronic amplifier.

They have the same advantage over thermal devices as flame ionisation types and are more easily applied because:

- flame contact is not required and positioning of the sensor is therefore simplified
- no electrodes are required
- there is no deterioration at high temperatures

The operation of the sensor depends on it receiving radiations from the flame. These initiate a flow of electrons which produces a signal current that may reach $100\,\mu$A. The current is amplified to operate a relay.

Although flames generally radiate over wavelengths from infra-red, through the visual range to ultra-violet, gas flames only produce weak visual radiations and flame sensors must therefore be sensitive to infra-red or ultra-violet rays.

Infra-Red Detectors*

The wavelength of the radiation sensed by this detector is usually from 1 to 3 μm. Because all heated objects emit infra-red radiation, a simple detector cannot differentiate between a flame and a hot refractory surface. However, the radiation from a refractory is relatively steady, whilst that from a flame modulates or "flickers". The detector is therefore designed to respond to a modulating radiation and not to a steady output. A lead sulphide cell is commonly used.

This cell can become "saturated" when exposed to high background temperatures and may also respond to a flicker effect caused by air passing over a hot surface. Infra-red sensors are therefore usually restricted to applications at temperatures below 800°C (1470°F).

Ultra-Violet Detectors

These detectors respond to radiations with wavelengths of 0.19 to 0.3 μm. At these wavelengths the energy emitted from a flame is

*This method of flame protection is no longer recommended.

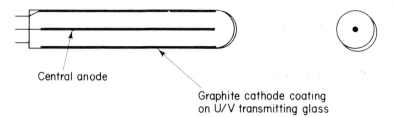

Central anode

Graphite cathode coating
on U/V transmitting glass

Fig. 10   Coaxial electrode (Geiger-Müller)

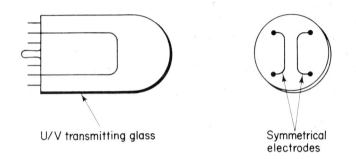

U/V transmitting glass

Symmetrical
electrodes

Fig. 11   Symmetrical diode

normally much higher than that from a hot or incandescent surface which radiates at $0.7\,\mu m$ or above. Sensors which operate in this narrow wave-band are therefore readily able to discriminate between the flame and surrounding refractory surfaces.

Early detectors used the Geiger-Müller tube whilst the type in common use is the symmetrical diode. Both types use a gas-filled ultra-violet transmitting quartz glass envelope. The G-M tube is shown in Fig. 10 and has a co-axial electrode structure consisting of a cathode of graphite material deposited on the inner surface of the glass and a central rod-like anode.

The symmetrical diode, Fig. 11 has two wire electrodes placed parallel to each other with a narrow space between. An alternating voltage from 220 to 900 V, depending on the type of unit, is applied across the electrodes. When ultra-violet radiations within the appropriate wave-band strike the electrode which is in the negative half-cycle, an electron is emitted. As this electron accelerates towards the positive electrode it strikes atoms of the inert gas which fills the tube, so ionising the gas. This results in an increased discharge between the electrodes and current flows. The current dies away as the voltage drops to zero but the discharge will be repeated when the voltage rises again, if UV radiation is still being received.

The a.c. produced by the tube may be rectified and amplified to operate a control relay.

UV detectors can be made self-checking by

- automatic shutters to cut off the radiation
- electronic cut-outs

An automatically controlled shutter is periodically interposed between the detector and the source of radiation to halt the discharge. If the current does not stop flowing, the burner is shut down and the system locks out.

Electronic cut-outs also periodically halt the discharge automatically. Both self-checking devices check for a false flame signal during the operation of the burner.

As an additional safeguard a safe-start check must be carried out.

The safe start check is made by applying a test voltage, considerably higher than normal operating voltage, to the detector during start up and pre-purge periods. If the detector does not discharge, the ignition sequence continues. If a discharge occurs the detector is faulty and the system goes to lock out.

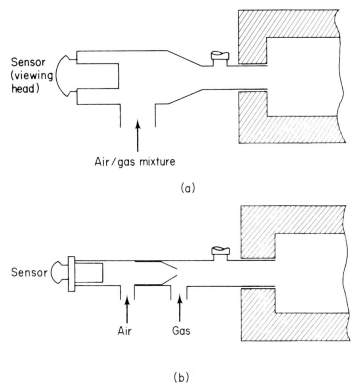

*Fig. 12 Location of ultraviolet flame sensor; (a) in air/gas mixture; (b) in combustion air. (Viewing in line of flame **not** recommended.)*

Installation

The satisfactory operation of photo sensitive detectors depends on:

- the correct mounting and wiring of the sensor
- the correct sighting of the sensor with respect to the pilot and main burner flames

Mounting should be carried out in accordance with the manufacturer's instructions. The tube should not reach temperatures of more than 50 to 75°C (122 to 167°F) depending on the type.

Where necessary the sensor can be located so that it is cooled by the combustion air or by the air/gas mixture, Fig. 12. Because fuel gases can absorb UV radiation from a flame the sensor should never normally be mounted to sight through a gas supply.

Wiring should be carefully installed with all joints sealed against moisture and leads from the sensor to the control box run separately and not in the same conduit as other wiring to avoid induced signals.

The length of the lead should normally not exceed 15 m (50 ft) and with some types should be considerably shorter. Some of the early symmetrical diode detectors had the flame relay mounted with the sensor in the viewing head.

Sighting is critical and the following points are important:

- the line of sight should ensure that the minimum pilot flame detected is of such a size and in such a position that it will always ignite the main burner
- UV detectors are sensitive to ignition sparks and a check should be made with all gas valves closed to ensure that the sparks are not detected

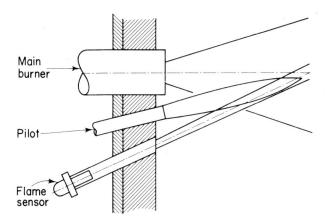

Fig. 13   Location of ultra-violet sensor (preferred method)

- although the UV sensor should never be sighted through fuel gas, it may view the flame through a stoichiometric air/gas mixture up to a distance of about 1 m
- for both infra-red and UV sensors the viewing tube should preferably be of black mild steel and not stainless or galvanised steel
- the line of sight should be such that the flames are viewed under all normal operating conditions; this may be checked by reading the voltage across the coil of the flame relay in accordance with the manufacturer's instructions

Commissioning

After carrying out normal checks a pilot turn down test should be applied as described in Section 91. If the main burner fails to ignite within 2 seconds, shut down immediately and resight the viewing head.

When checking shut down, the safety shut off valve should close within 1 second.

Servicing

This should be carried out regularly and frequent checks should be made to ensure that the system will operate satisfactorily and go to lock out if a hazardous situation occurs.

A UV cell usually has a maximum life of 10,000 hours or 1 year at $50°C$ ($122°F$). Although the most common fault is for a cell to fail to detect a flame which is present the cell can "go soft" and detect a flame which is not present. This is a failure to danger, but it should be detected when the cell is checked each time the burner is started up.

Automatic programming control units incorporate a safe-start check and if the cell senses a flame in the pre-purge period the system goes to lock out.

Semi-automatic units prevent the pilot valve and ignition transformer being energised if the cell has gone soft.

The more frequently units are started up, the more often a check is made on the UV cell. If not shut down daily by a time clock, the system must be shut down manually at least once per day.

Large burner plant is frequently fitted with a self-checking system. Every few seconds the flame is obscured electrically or mechanically from the UV cell. If the system does not start to shut down it is overridden and goes to lock out.

§94 Automatic and Semi-Automatic Controls

The basic differences between manual, semi-automatic and automatic control are shown pictorially and by means of a table in Fig. 14.

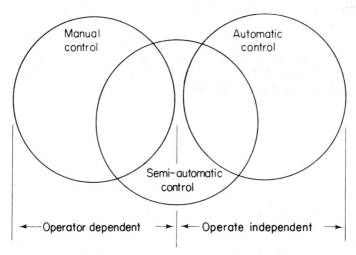

Stages	Time limits for each stage		
	Manual	Semi-automatic	Automatic
Pilot ignition	—	Unlimited (or preset)	$3\frac{1}{2} \pm 1\frac{1}{2}$ S
Pilot proving	—	None	> 5 S
Main flame establishment	Unlimited	Unlimited (or preset)	$3\frac{1}{2} \pm 1\frac{1}{2}$ S
Flame supervision	Varied	Continuous on pilot (and main flame)	Continuous on pilot and main flame

Fig. 14 Comparison of manual, semi-automatic and automatic control

Definitions of automatic and semi-automatic controls are as follows:

Automatic Control

An automatic control system is one which can establish a main flame from the completely shut down condition without any manual intervention. So the main burner may be controlled automatically by a thermostat, timer or similar switching device.

The flame on an automatic forced draught burner is established in stages as follows:

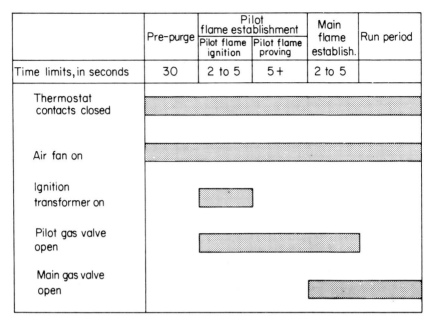

	Pre-purge	Pilot flame establishment		Main flame establish.	Run period
		Pilot flame ignition	Pilot flame proving		
Time limits, in seconds	30	2 to 5	5+	2 to 5	
Thermostat contacts closed					
Air fan on					
Ignition transformer on					
Pilot gas valve open					
Main gas valve open					

Fig. 15   Automatic control unit sequence diagram (on-off operation)

- pre purge   —combustion air fan on to clear any combustibles which may have accumulated in the combustion chamber
- pilot ignition   —pilot gas and ignition spark both on
- pilot proving   —detection of stable pilot by flame safeguard with ignition spark off
- main flame ignition—pilot gas and main gas supply on
- main flame proving —detection of stable main flame with pilot gas off (if interrupted pilot is used)

On shut-down, which could be due to the thermostat contacts opening, the programmer returns to the beginning of the sequence in readiness for the thermostat contacts to close again. Figure 15 shows the time interval for each stage of the sequence.

Semi-Automatic Control

The opposite to an automatic control is a manual system where the operator carries out each stage of the sequence. The semi-automatic system incorporates elements of the two extremes. It usually provides spark ignition and flame detection but always requires some manual action to bring on the main burner from the shut down condition. This may simply involve releasing a spring loaded start button after the pilot has ignited.

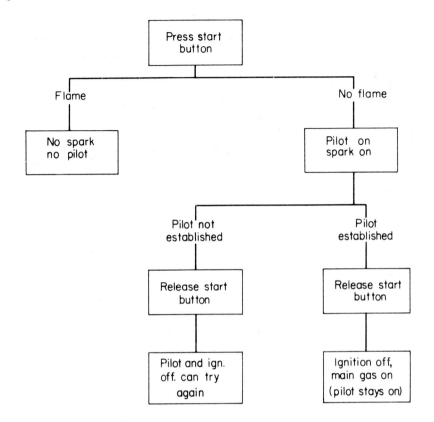

Fig. 16   Typical light-up sequence for semi-automatic control unit

A typical semi-automatic control sequence is:

- pilot ignition   —operator presses start button, pilot gas and ignition spark on
- pilot proving/main flame ignition   —operator releases start button, pilot gas and main gas on, provided that pilot flame has been detected.

The sequence is shown as a chart in Fig. 16.

With both automatic and semi-automatic control, combustion safeguards, such as ultra violet and flame rectification detectors and supply safeguards, for example pressure switches, are incorporated. These devices give shut down or lock out, requiring manual resetting when a fault condition occurs. Figure 17 shows layouts of controls, including flame safeguard, for natural draught and forced draught burners.

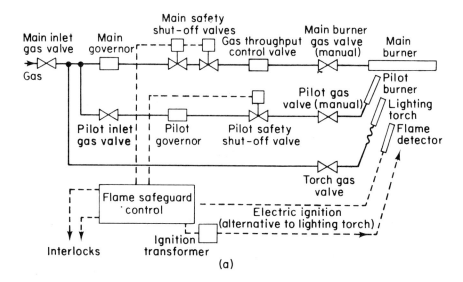

(a)

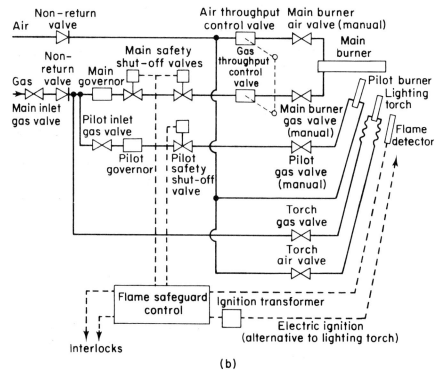

(b)

Fig. 17   Controls, including safeguards, on (a) natural draught, and (b) forced draught burners

Commercial Catering

Chapter 6 is based on an original draft by Mr. T. Fox

§95 Introduction

There are many different catering operations carried out in a wide variety of establishments. Food is served in pubs and clubs, hotels and hospitals, shops and schools, cafés and canteens.

Nearly half of all the meals served in a year by catering establishments are produced in hotels and guest houses, cafés, snack bars and restaurants, public houses and clubs, fish and chip shops and take-aways.

About one quarter are produced in schools and colleges, while hospitals, office and industrial canteens together produce about one tenth.

In addition to the establishments which produce meals or snacks for consumption on the premises or to take away, there are others which process food for sale in shops. These are the bakers, confectioners and the makers of pies, biscuits and frozen foods. Where these operations are carried out by the large combined organisations, the plant is mechanised and the kitchen becomes a factory.

This chapter deals generally with the smaller scale unautomated commercial cooking operations.

§96 Cooking Processes

The basic cooking processes and the domestic appliances on which they may be carried out were described in Volume 2, Chapter 6. The information given applies equally to those catering appliances which are simply larger and more robust versions of domestic appliances. For large scale cooking operations specialised equipment is used and the cooking process may be modified.

Baking

Baking is cooking food dry in an oven, principally by convection with some radiation from the oven walls. Cooking times may be reduced by using forced convection. Bread is best baked on a falling temperature from 260°C (500°F) down to 230°C (450°F) and steam may be introduced into the oven for a period to give a crisp brown crust to the products (see Chapter 8).

Blanching

This is carried out to remove skins from fruit or nuts. Vegetables may be blanched before being frozen for storage to check the action of the enzymes which cause deterioration.

Blanching is also carried out to whiten food or to remove strong flavours. It is usually done by placing the food in a wire basket and immersing it in boiling water for 2 to 5 minutes. This is immediately followed by rapid cooling in cold water.

Boiling

This is heating food by immersing it completely in a liquid, essentially water, at 100°C (212°F). Water is boiling when bubbles continually rise to the surface. Because some evaporation takes place more liquid may need to be added during boiling.

Meat, poultry and green vegetables are added to boiling water which is quickly brought back to the boil.

Root vegetables, except new potatoes, are placed in cold water and then brought to the boil.

Any scum arising from boiling must be removed. With the exception of green vegetables, food may be covered during cooking.

Braising

Braising is slow cooking in a tightly covered container with some liquid or sauce, usually in an oven. Meat and poultry are browned by frying in a little fat before being braised. They are then laid on a bed of sliced vegetables and half covered with liquid. Fish and vegetables may also be braised. In the case of meat and fish, the braising liquid is used to make the accompanying sauce.

Frying

There are various methods of frying:

- deep frying
- shallow frying
- dry frying

Deep frying is done by totally submerging the food in hot fat or oil. With the exception of potatoes, the items are usually coated with batter, seasoned flour or egg and breadcrumbs before frying. Potatoes are dried thoroughly. Normal frying temperatures are 185 to 195°C (365 to 383°F).

Shallow frying is cooking in a small amount of fat in a shallow pan. The food is cooked on the best side first and turned so that the fat seals the whole of the outside. The items should be well drained after cooking. "Sauté" is a term given to small items of food which are tossed while being shallow fried. "Meuniere" is the shallow frying of fish, usually in butter.

Dry frying is carried out on a lightly greased flat plate and is used for cooking eggs, bacon, liver, hamburgers and pancakes. The plate or griddle is tilted to allow any fat to run to a drain so that the food stays reasonably dry.

Grilling

This is cooking by radiant heat. The food is placed on bars and in the original grills, was heated from below. Grills may now be "underfired" or "overfired". Overfired grills or salamanders have their heat source above the food while underfired or "flare" grills have theirs below. In the latter, the liquid fat from the meat falls on the hot fuel and flares up. The smoke and flames give the meat a characteristic flavour and appearance.

Poaching

Poaching is heating food totally submerged in water just below boiling point at a temperature of about 93 to 95°C (200 to 203°F). This is as close to boiling point as possible without there being any movement in the water. Poaching is used for cooking fish, fruit and eggs.

Poeler

This is cooking meat or poultry in a covered container with butter, in an oven. The meat may be laid on a bed of sliced vegetables and there must be sufficient butter for basting. The lid is removed to complete the cooking and the liquid is used to make the accompanying sauce.

Roasting

Roasting may be carried out in an oven or on a rotating spit. The food should be cooked with a little fat and must be basted occasionally to keep it moist. Joints of meat and poultry should be seasoned before cooking and should preferably be raised off the bottom of the roasting tin. Roasting should begin at a high temperature to seal in the meat juices. The temperature may then be reduced. Vegetables should be partially cooked before being put in the oven.

Simmering

Simmering is very gentle boiling or cooking at a temperature just below boiling point but above poaching temperature. The liquid should just show slight movement. Soups and stock are simmered gently. The surface must be skimmed periodically to remove any impurities.

Steaming

This is cooking food by enclosing it in a compartment filled with steam from boiling water. Steaming may be carried out at atmospheric pressure or at higher pressures from 35 mbar to 1 bar (0.5 to 14.5 lbf/m²). At the higher pressures, temperatures are raised and cooking times are reduced.

The method is used for puddings and vegetables which can be cooked slowly without losing their colour or flavour.

Stewing

This is the slow cooking of meat and vegetables in enough stock or sauce to just cover the food. Gentle simmering concentrates the flavour and tenderises the tougher cuts of meat. Casseroles and hot-pot are forms of stew which are baked in the oven.

§97 Equipment

Gas fired catering equipment is designed to comply with the recommendations of BS 2512.

The type of equipment used in an establishment depends on the following factors:

- the standard of catering
- the menu offered, number of courses, choice of dishes
- type of service, cafeteria, table service either plated or waiter served
- number of meals provided
- type of food used, fresh, frozen or convenience.

Small cafes and boarding houses may be able to manage with only a single-oven range. The large five-star hotels with specialist chefs for meat, fish, vegetables, soups, pastries and so on will use the complete range of specialised equipment which follows.

§98 Ranges

A range consists of a boiling table or hotplate mounted on a general purpose oven. An eye level grill may be incorporated into the potrack.

Ranges may be classified as either medium or heavy-duty. Medium-

Fig. 1 Medium duty range

duty ranges usually have open-top hotplates and internally heated ovens with single or double doors, hinged at the side. They may incorporate a grill and they are, in essence, a larger and more robust version of the domestic cooker, Fig. 1.

Heavy-duty ranges usually have solid tops and semi-externally heated ovens with drop down doors. They are designed to withstand continuous use with heavy cooking pans, Fig. 2.

Hotplates

Open top hotplates, Fig. 3, may have various designs of ring burners giving immediate flame contact with the cooking utensil.

Solid top hotplates, Fig. 4, are cast iron and are ribbed on the underside. They are usually heated by a central burner surrounded by refractory brick. The "bulls-eye" and its surrounding rings may be removed to allow flame contact if required. Temperatures on the top range from 540°C (1000°F) at the centre to 150°C (300°F) near the edge.

Fig. 2   Heavy duty range

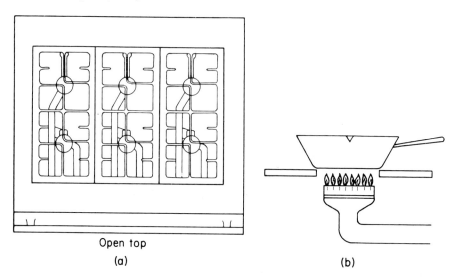

Open top
(a)

(b)

Fig. 3   Open top hotplate (a) plan view (b) burner

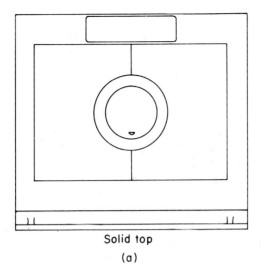

Solid top

(a)

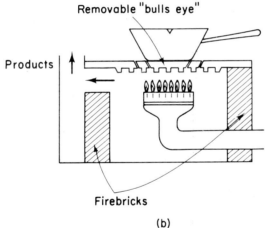

Fig. 4 *Solid top hotplate (a) plan view (b) burner*

Ignition is usually by permanent pilot or spark igniter and solid top burners have flame protection devices.

Gas rates may be as follows:

- open top   3 to   6 kW (10 000 to 20 500 Btu/h)
- solid top 12 to 15 kW (41 000 to 51 000 Btu/h)

Ovens

The three basic types of oven are:

- internally heated or direct, Fig. 5
- semi-externally heated, Fig. 6
- externally heated or indirect, Fig. 7

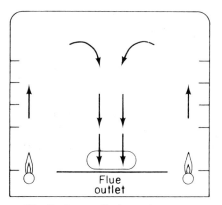

Fig. 5   Internally heated oven

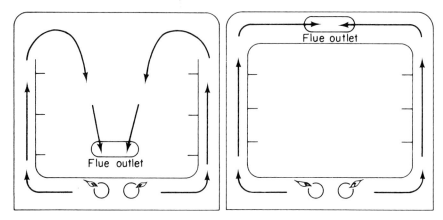

Fig. 6   Semi-externally heated oven   *Fig. 7   Externally heated oven*

Internally heated ovens have a temperature gradient which allows different foods to be cooked at the same time, while externally heated types have an even temperature throughout.

Ranges usually have one of the first two types and the fully indirect design is commonly used as an independent baking oven.

Ignition may be by spark or permanent pilot and flame protection devices are incorporated. Most ovens have thermostats.

Gas rates are from 7 to 9 kW (24 000 to 31 000 Btu/h).

§99   Boiling Tables and Stockpot Stoves

Boiling Tables

These are hotplates mounted on low tables with a working height of 760 to 860 mm (30 to 34 in). They may have open or solid tops.

Like ranges they are designed for either medium or heavy duty.

Open top boiling tables have various types of burners and often include double or treble concentric burners. The solid tops are heated by single or multiple ring burners or jet burners.

Boiling tables are usually fitted with a steel storage shelf and may have a splash plate and pot racks, Fig. 8. Ignition is by spark or permanent pilot and solid tops have flame failure devices.

Gas rates may be from 4 to 7 kW (13 500 to 24 000 Btu/h) for open top burners and up to 19 kW (65 000 Btu/h) for solid tops.

Stockpot Stoves

A stockpot stand or stove is a low level boiling table designed to heat one large heavy utensil. The working height is usually 450 to 600 mm

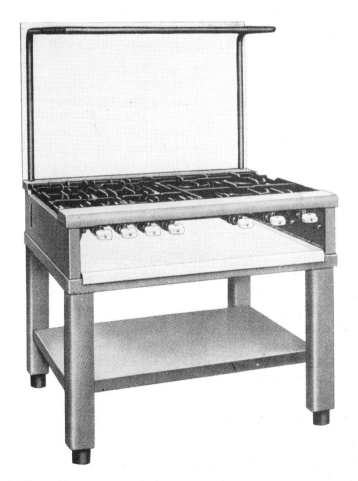

Fig. 8 Boiling table

Fig. 9   Stockpot stoves (a) single burner (b) double concentric ring burner

(18 to 24 in). It may be heated by a single ring burner or by concentric rings, Fig. 9. Gas rates are from 8 to 12 kW (27 000 to 41 000 Btu/h).

§100   Ovens

Ovens may be classified as follows:

- general purpose and roasting ovens
- forced convection ovens
- pastry ovens
- proving ovens
- large scale baking ovens

General Purpose and Roasting Ovens

These may be part of a range or separate units, either as single ovens or tiered, Fig. 10. They are generally either internally or semi-externally heated and heavy duty ovens usually have drop down doors.

Ignition is by spark or pilot with flame protection. Thermostats may be bimetal or liquid expansion and could be direct, or indirect operating through a relay valve.

Gas rates may be:

- general purpose ovens, 7 to 10 kW (24 000 to 34 000 Btu/h)
- roasting ovens, 15 to 16 kW (51 000 to 54 500 Btu/h)

Fig. 10   General purpose oven

Forced Convection Ovens

These ovens have the heated air circulated throughout the interior by means of an electric fan. Figure 11 shows a two-tier model. They may be either externally or semi-externally heated and can be used for all normal roasting and baking operations. The ovens have an even heat distribution throughout and heat up quickly, giving a high output. They are particularly suitable for reheating or end cooking frozen foods.

Semi-externally heated ovens, Fig. 12, have some of the combustion products mixed with the circulating air. Air from the fan is deflected upwards through venturi shaped channels, into the side chambers. The reduced pressure at the throat draws in the combustion products. These pass with the hot air through the perforated side panels and around the oven. Some hot gases escape through the flue outlet and the remainder are drawn through the perforated rear panel downwards to re-enter the fan.

Externally heated ovens, Fig. 13, heat the hot air indirectly and the combustion products pass around the oven to the flue outlet.

Ignition is usually by an automatic mains operated spark igniter with mercury vapour flame failure device.

Fig. 11   Forced convection oven

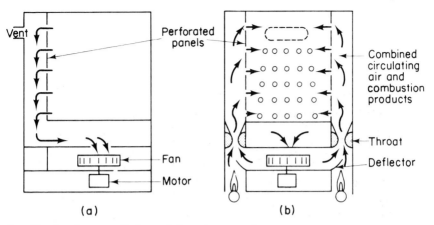

Fig. 12   Semi-externally heated forced convection oven

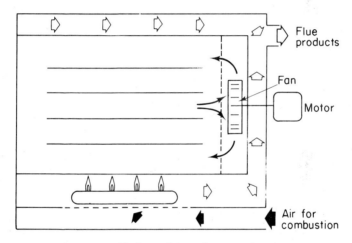

Fig. 13   Externally heated forced convection oven

Other controls may include:

- thermostat
- door switch to shut off fan and main gas when doors are opened
- auxiliary fan switch for rapid cooling
- load control to adjust the heat input to suit the number of racks in use

Timers are usually included.

Gas rates vary considerably from the small counter top models at about 5 kW (17 000 Btu/h), up to 32 kW (109 000 Btu/h) for a large, freestanding oven.

Pastry Ovens

These are externally heated, low height ovens designed to have an even temperature distribution. They are used for baking cakes, pastries and tarts of various kinds. The ovens are from 125 to 300 mm (5 to 12 in) in height and cooking is usually done on the sole plate in those below 200 mm (8 in). Larger ovens may have a shelf.

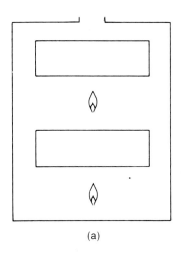

(a)

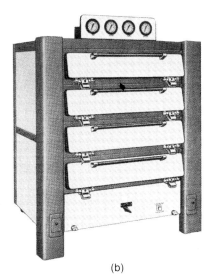

(b)

Fig. 14   Pastry oven (a) decks heated by individual burners (b) single burner system

Because of their low height, pastry ovens are usually grouped in ties of two, three or four ovens. Each deck may be heated separately or all decks may have a common burner system, Fig. 14. In the second case, separate thermometers indicate the differences in temperatures between the ovens.

The atmosphere in the ovens can be controlled by a vent in the door. This gives a steamy atmosphere when closed and a drier heat when opened.

Ignition is usually by permanent pilot and flame failure devices are fitted. Thermostats are generally indirect, with relay valves. Thermometers are often included.

Gas rates may be from 9 to 14.5 kW (31 000 to 49 500 Btu/h).

Proving ovens

When yeast is mixed with sugar and water in a warm atmosphere it ferments and gives off carbon dioxide. Mixed with dough the yeast produces little bubbles which aerate the mixture. This fermentation is called "proving" and it must be carried out in a warm, moist atmosphere to make the dough rise before cooking. Proving ovens are used for this purpose.

The ovens are large, low temperature ovens with water pans to produce vapour and with water feed tanks. They maintain temperatures of 26 to 32°C (79 to 90°F) by means of a thermostat.

Gas rates are about 2 kW (6800 Btu/h)

Baking Ovens

There are various types of baking ovens still in use including:

· steam tube ovens
· forced convection peel ovens
· rack ovens
· reel ovens
· continuous hearth conveyor ovens

Steam tube ovens are heated by heavy gauge, small bore wrought iron tubes running through either the roof or the base or "sole" of each deck. The tubes are sealed and contain a small quantity of distilled water which turns into high pressure steam when the tube ends are heated, Fig. 15.

In addition to the usual gas controls, the oven must have a manually reset overheat cut off.

Forced circulation peel ovens are named after the tool which is used to load them. A "peel" is a long wooden pole with a flat, spade shaped end used for charging and discharging any fixed oven. The ovens are heated externally by forced circulation of the combustion products giving even heat distribution and quick heating up.

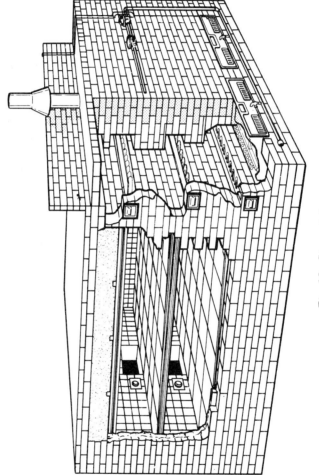

Fig. 15 Steam tube oven

Fig. 16 Rack oven

Rack ovens are forced convection ovens, usually externally heated. The bread or other goods to be baked are loaded on to trays which are placed in a rack which is pushed into the oven, Fig. 16. There are various methods by which the racks may be rotated or moved round the oven. Rack ovens are ideal for batch production.

Reel ovens are a form of conveyor oven in which the products are placed on trays suspended from a large rotating reel, similar to a

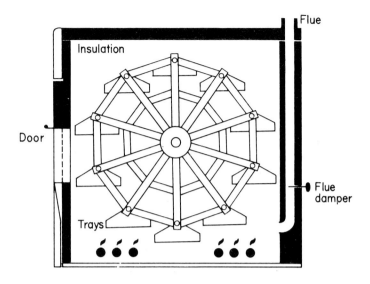

Fig. 17   Reel oven

ferris wheel at a fair ground, Fig. 17. The oven is usually internally heated and the reel is geared so that articles return to the oven door every $2\frac{1}{2}$ minutes. This enables different products to be cooked at the same time and removed when desired. Reel ovens may also be used for roasting.

Ignition is by mains operated spark and the temperature is controlled by an electrical thermostat, solenoid valve and relay valve. Flame failure devices are included.

Gas rates vary widely from about 13.5 to 95 kW (46 000 to 324 000 Btu/h). Continuous hearth conveyor ovens were described in Chapter 3. They are used in the major plant bakeries and are run at high temperatures with controlled steam injection. This produces a moist loaf lacking the flavour achieved in a smaller oven by the local baker.

§101   Grillers

Underfired, or flare grills are commonly used for steaks and chops and are often in public view. The source of heat is below the food and consists of a burner firing upwards on to a bed of refractory material or pumice, Fig. 18(a). Overfired grills, or salamanders can, in addition, be used for making toast and salamandering. They have the heat source above the food, Fig. 18(b). This may comprise sets of

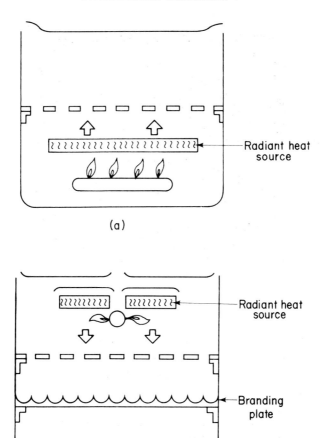

Fig. 18   Griller (a) underfired (b) overfired

burners firing below refractory or metal frets, or surface combustion plaques. A "branding plate" is usually supplied. This is a fluted, solid aluminium alloy plate which absorbs heat so that food placed on it, below the frets, is cooked on both sides at the same time. Figure 19 shows an overfired grill with a branding plate.

Ignition on grillers is usually manual or by permanent pilot.

Gas rates range from 5 to 16 kW (17 000 to 54 500 Btu/h for overfired grills and up to 30 kW (102 500 Btu/h) for underfired models.

Fig. 19   Overfired grill with branding plate

§102   Griddle Plates

Griddles are solid metal plates heated from below by burners to a temperature of 200 to 300°C (390 to 570°F) Fig. 20(a) & (b). The plate may have a channel to drain off the fat from cooking. They are

(a)

Fig. 20(a)   Griddle plate, counter model

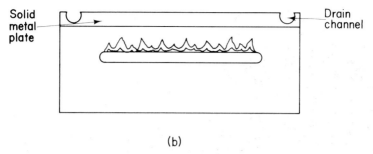

(b)

Fig. 20(b) Griddle plate, cross-section

used for dry frying of eggs, bacon and similar foods or for pancakes or drop scones. Sizes vary from small counter top units to larger free-standing models. They are commonly used for "call order" cooking in snack bars.

Ignition is usually manual and gas rates vary from 5 to 15 kW (17 000 to 50 000 Btu/h).

03 Fryers

There are two main types of fryer:

- deep fat fryer
- tilting fryers or brat pans

Deep Fat Fryers

These are heated pans containing oil or fat. They are used for cooking chips, fish, poultry, sausages, onion rings, fritters and doughnuts, Fig. 21. There are three types of deep fat fryers:

- flat bottomed
- 'V' pan
- immersion tube

Flat bottomed fryers, Fig. 22, have a flat pan, finned or studded below and heated directly by a burner. This type is used in fish and chip shops.

'V' pan fryers have a vee-shaped pan, heated above the bottom of the vee, Fig. 23. Immersion tube fryers are heated by burners firing through tubes which pass through the pan from one side to the other, Fig. 24.

In both the latter types there is a cool zone below the burners into which food particles can sink and remain without charring. This prevents the particles transmitting flavours from a previous load to

Fig. 21   Deep fat fryer

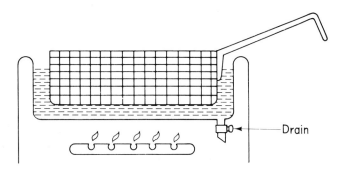

Fig. 22   Flat-bottomed fryer

the new food being cooked. In flat pans contamination can only be avoided by straining out the solid particles after every frying.

Frying temperatures are critical and must be maintained with close limits. Different fats and oils have different maximum temperatures or "smoke points". At or above its smoke point the flavour of food will be spoiled and above the smoke point spontaneous combustion can occur. Table 1 (§104, page 220) gives the smoke points of common cooking fats.

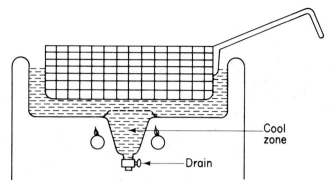

Fig. 23   'V' pan fryer

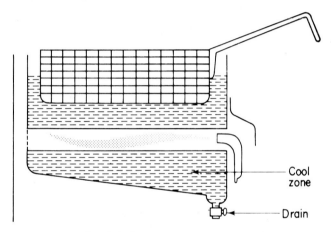

Fig. 24   Immersion tube fryer

Ignition is by manually lit pilot with flame protection. Thermostatic control is usually by an indirect thermostat and relay valve. The thermostat is adjustable to a maximum temperature of 190°C (374°F). Hospital management boards require an additional thermostat or over-heat cut off set to shut down at 215°C (419°F). Timers may be included.

Gas rates vary from about 8 kW (27 500 Btu/h) for small counter top models up to 35 kW (119 500 Btu/h) for large free standing models.

Tilting Fryers or Brat Pans

These are shallow, flat bottomed pans capable of being tilted forward by a hand wheel or a lever, Fig. 25. They can be used for shallow or dry frying and also for boiling, poaching or stewing.

§104   TABLE 1   Smoke Point of Common Fats and Oils

Cooking Medium	Smoke Point	
	°C	°F
Coconut oil	138	280
Ground nut oil	149–243	300–469
Dripping	163	325
Olive oil	169	336
Maize oil	221	430
Lard	190–221	374–430
Butter fat	208	406
Cotton seed oil	233	450

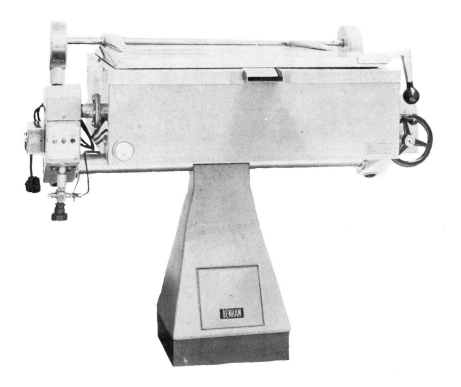

Fig. 25   Tilting fryer or brat pan

Ignition is usually by permanent pilot with flame protection device. A control can prevent the pan from being tilted while the main gas is on.

Gas rates are 12 to 18 kW (41 000 to 61 500 Btu/h).

05 Steaming Ovens

Steaming ovens may be either atmospheric or pressure types. Atmospheric types are designed for either light or heavy duty. Figure 26 shows a light duty atmospheric model. Steaming ovens are used for cooking root vegetables, fish or puddings.

The ovens are heated by steam from a water trough in the base which has gas burners below it.

Fig. 26   Light duty atmospheric steaming oven

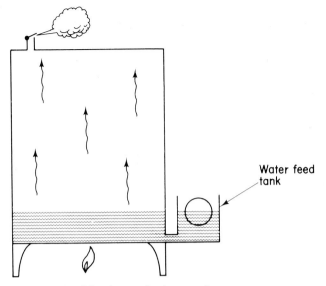

Fig. 27   Atmospheric steaming oven

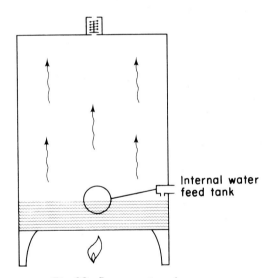

Fig. 28   Pressure steaming oven

Atmospheric steamers have their water supplied by an external cistern and steam is allowed to escape through a vent at the top of the oven, Fig. 27. Ignition may be manual or by a permanent pilot with flame protection. Temperature is controlled by an indirect thermostat and relay valve.

Pressure steamers have an internal ball valve and a spring loaded pressure relief valve in place of the vent, Fig. 28. Ignition is usually by protected permanent pilot and control may be by a pressurestat. Most pressure types operate at 35 mbar ($\frac{1}{2}$ lbf/in^2) but models working at pressures of about 1 bar (14.5 lbf/in^2) are available. The high pressure models are used for cooking frozen foods.

Gas rates vary and may be from 13 to 20 kW (44 500 to 68 000 Btu/h).

106 Boiling Pans

Boiling pans are large-capacity vessels holding from 45 to 180 litres (10 to 40 gallons) and used for boiling food in bulk. There are three main types:

Fig. 29 Jacketed boiling pan

- single pan
- jacketed pan
- dual purpose

A jacketed pan is shown in Fig. 29.

Single Pan

The pan is heated directly by the burner, Fig. 30. A large bore draw off tap is fitted at the base of the pan so that the liquid may be drained off. Pans are fitted with lids which may be pivoted and

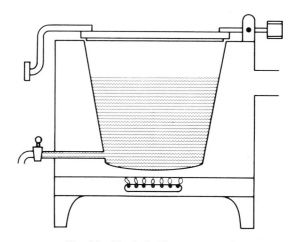

Fig. 30   Single boiling pan, section

counter balanced on large models. The food is sometimes held in wire baskets to assist handling and prevent it sticking to the sides of the pan. Single pans are used for cooking vegetables or meat and making soup.

Jacketed Pans

The pan is heated indirectly by hot water or steam in an outer pan, Fig. 31. This gives   an even heat distribution and avoids local over-heating and food sticking or burning. Jacketed pans are used for cooking thick soups, porridge, custard or milk puddings.

A water cock with a swivel arm is usually fitted which supplies water directly to the inner pan and, on some models, through a funnel to the outer pan. There is usually a thermally operated air vent between the funnel and the pan. This is open when cold to allow the pan to fill up and closes when heated to pressurise the pan to about 500 mbar (7 lbf/in^2). These pans are fitted with a safety valve and a water gauge.

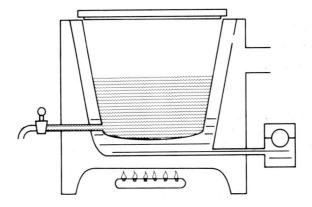

Fig. 31   Jacketed pan, section

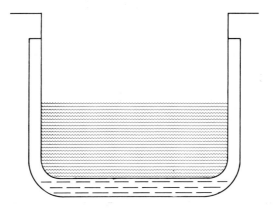

Fig. 32   Pan with sealed water jacket

The pan shown in Fig. 31 has the outer pan supplied by an external cistern with a ball valve.

Some models have a completely sealed water jacket, Fig. 32. They are fitted with a bursting disc to relieve excessive pressure.

Dual Purpose Pans

There are single pans which have an inner removable pan or porringer and can be used either as a single pan or as a jacketed pan, Fig. 33.

Ignition is usually by permanent pilot, sometimes with flash tube. Flame failure devices are generally fitted. On sealed jacket pans the controls may include pressurestat, relay valve, safety valve, air vent, pressure gauge and multi-functional controls. The safety valve operates at about 750 mbar (11 lbf/in^2) and gas rates vary from 9 to 35 kW (31 000 to 119 500 Btu/h).

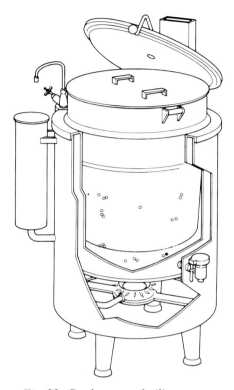

Fig. 33   Dual purpose boiling pan

§107   Hot Cupboards

These are insulated cabinets which are used to heat plates or to keep cooked food hot, Fig. 34. There are two main types:

- directly heated
- indirectly heated

Direct types obtain their heat from a burner situated below a baffle plate, Fig. 35. They are generally used for plate warming at about 60°C (140°F).

Indirectly heated models are heated either by hot gases circulated around the cabinet through channels, or by steam generated in a water trough in the base, Fig. 36. They have a more humid atmosphere and are suitable for keeping food hot at about 82°C (180°F).

Ignition is usually manual and hot cupboards are usually thermostatically controlled to pre-set temperatures. Gas rates range from about 2 to 7 kW (7000 to 24 000 Btu/h).

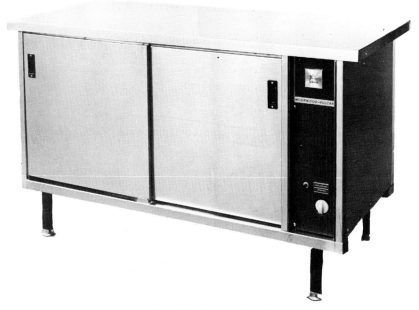

Fig. 34   Hot cupboard

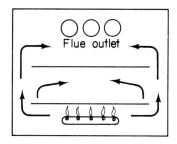

Fig. 35   Directly heated hot cupboard, section

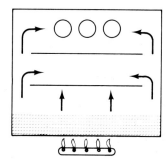

Fig. 36   Indirectly heated hot cupboard, section

§108 Bains Marie

"Bain" is the French word for "bath" and bains marie are appliances used to keep foods hot before or during serving. In their simplest form they consist of an open trough containing water heated from below by a gas burner, Fig. 37. The pans or vessels containing food are stood on a rack in the water.

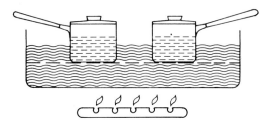

Fig. 37 Open trough bain marie

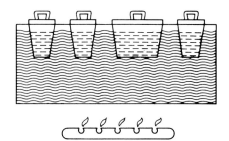

Fig. 38 Wet heat fitted bain marie

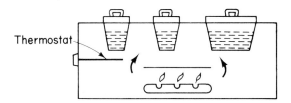

Fig. 39 Dry heat fitted bain marie

Fitted models are available in which the water is covered by a "filler plate" into which are set containers of various shapes and sizes to hold different kinds of food, Fig. 38. Some fitted models are directly heated by a burner below a baffle plate, Fig. 39. The temperature of the food is maintained between 75 and 85°C (167 and 185°F).

Ignition is usually manual with interlocking main and pilot taps, thermostats are normally fitted to directly heated types.

Gas rates vary from 5 to 11 kW (17 000 to 37 500 Btu/h).

109   Dish Washing Equipment

This comprises:

- sterilising sinks
- dish washing machines

Sterilising Sinks

There are stainless steel sinks heated from below by a gas burner, Fig. 40. They operate at 82 to 88°C (180 to 190°F). After washing up, the crockery and cutlery is loaded into racks and immersed in the sink. The utensils retain sufficient heat to dry out rapidly when

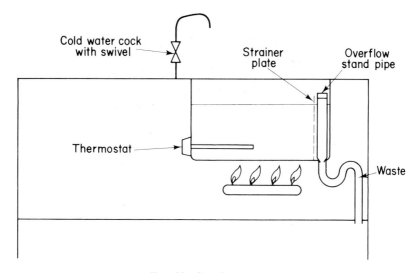

Fig. 40   Sterilising sink

removed, so eliminating the need for drying cloths. The sinks may be part of a washing up unit, Fig. 41. The units are used in small kitchens as the sole means of washing up or as stand-by units in larger establishments. Balanced flue models are available.

Ignition is by pilot with interlocking taps or flame failure device. Temperature is controlled by an indirect thermostat and a relay valve.

Gas rates may be 8 to 16 kW (27 500 to 54 500 Btu/h).

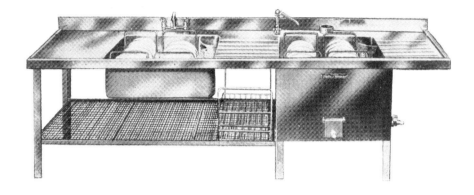

Fig. 41   Washing up unit with sterilising sink

Dish Washing Machines

There are basically two main types of dishwasher:

· front loading, under counter models
· conveyor types

The front loading models are similar to the domestic washers with drop down doors and pull-out racks. The conveyor washers may be manual, semi-automatic or fully automatic. Most machines have time washing cycles and automatic detergent dispensers. They are designed to deal with cutlery, crockery and glass ware. Special machines are available for pot washing.

Dishwashers are usually plumbed into a hot water supply from a separate boiler. Some have integral burners to raise the water temperature to about 82°C (180°F) for the final rinse.

§110   Bulk Water Boilers

Bulk water boilers are cylindrical containers heated by gas burners. They are generally constructed of:

· heavy gauge copper, tinned internally
· stainless steel with a tinned copper, pyrex or stainless steel liner

There are three main types:

- urns
- jacketed urns
- large boilers

Urns

These are simple containers with a lid. They have a draw off tap near to the base and may have a thermometer to indicate the temperature and a gauge glass to show the liquid level, Fig. 42. Urns are used to produce boiling water for tea or coffee making in quantities from 9 to 45 litres (2 to 10 gallons).

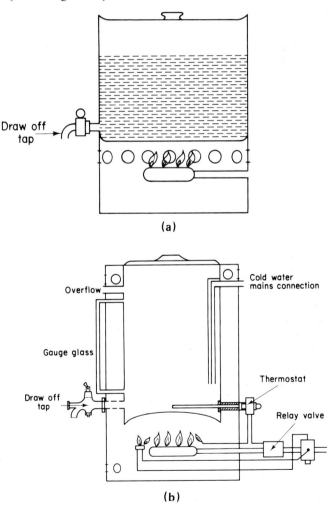

Fig. 42 *Bulk water heating urns*

Jacketed Urns

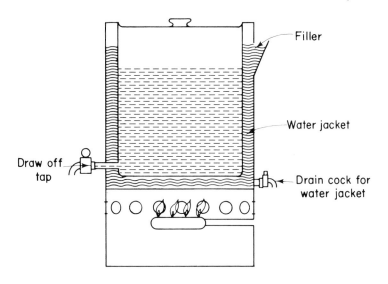

Fig. 43   Jacketed urn

Heated indirectly by a water jacket to prevent local overheating, these are used for milk or similar liquids, Fig. 43.

Large Boilers

These are similar to the urns but are fitted with cold water mains and overflow connections, Fig. 44. They are used particularly where large quantities have to be drawn off quickly. Capacities may be up to 136 litres (30 gallons).

Ignition of the simple models is manual. Larger boilers may have automatic ignition with a flame failure device and temperature control by thermostat and relay valve.

Gas rates range from 8 to 17 kW (27 500 to 58 000 Btu/h).

§111   Café Boilers

These are automatically operated water boilers which give continuous outputs of boiling water for use within about four minutes of lighting up. Side urns may be fitted for heating and storing milk and for

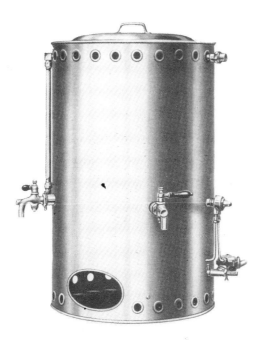

Fig. 44   Large water boiler

making and storing coffee. A combined appliance with side urns is called a "café set".

There are two main types of boiler:

- expansion boilers
- pressure boilers

Expansion boilers

These rely for their operation on the fact that water expands when heated. They are designed with the draw-off tap slightly higher than the water level in the boiler. Only water which is forced up into the expansion chamber by boiling can be drawn off, so only freshly boiled water is delivered, Fig. 45. The water supply is fed automatically from an external cistern with a ball valve, located at the rear of the boiler, Fig. 46.

Ignition is usually by swing-in pilot and the simplest boilers have interlocking taps and a gas tap linked with the draw-off tap. More complex models have automatic control devices and may store a reserve of hot water for any sudden demand.

Gas rates may be from 12 to 36 kW (41 000 to 123 000 Btu/h).

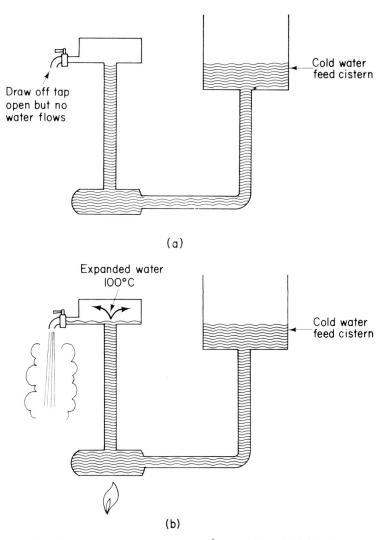

Draw off tap
open but no
water flows

Cold water
feed cistern

(a)

Expanded water
100°C

Cold water
feed cistern

(b)

Fig. 45   Principle of expansion café boiler (a) cold (b) boiling

Pressure Boilers

Pressure boilers are usually housed below a counter with the draw-off mounted above, Fig. 47 (page 236). Boiling water is forced up to the drawoff by the pressure of the steam produced, so water can only be drawn off when it is boiling, Fig. 48 (page 237). Some models include a steam drawoff or injector for heating liquids almost instantly. The boilers operate at pressures of 300 to 520 mbar (4.5 to

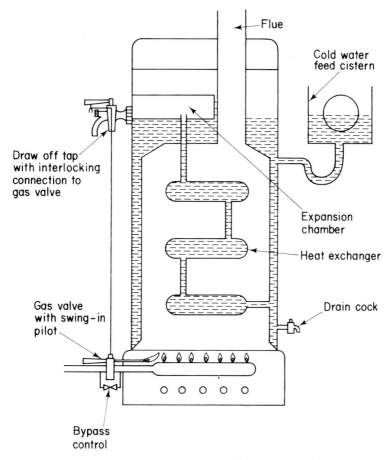

Fig. 46   Expansion type café boiler, section

7.5 lbf/in^2) and the temperature of the stored water is about 105°C (220°F).

Ignition is by permanent pilot with flame protection.

Other controls are usually:

- pressurestat
- water regulator
- safety valve
- excess temperature cut off

The pressurestat, Fig. 49 (see page 237), is a direct acting type. It consists of a bellows A, mounted on a diaphragm B, with its bottom face forming a gas valve C. The bellows is connected to the steam pressure and operates against the force of a spring D. The tension

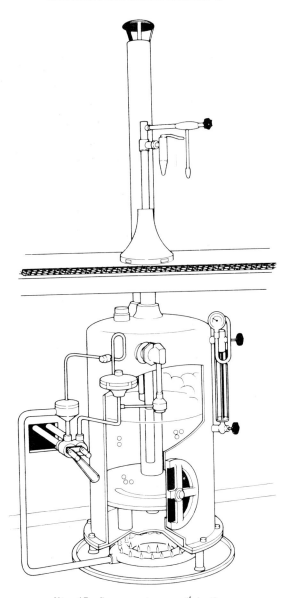

Fig. 47   Pressure type café boiler

of the spring determines the pressure at which the boiler will shut down to a bypass rate.

The water regulator, Fig. 50 (page 238), ensures that incoming cold water does not reduce the steam pressure below about 170 mbar (2.5 lbf/in²). It consists of a stout rubber diaphragm A, to which is attached a water valve B. Steam pressure on the diaphragm can open

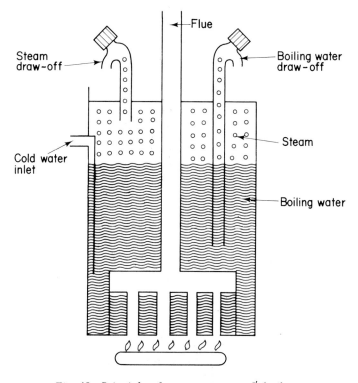

Fig. 48   Principle of pressure-type café boiler

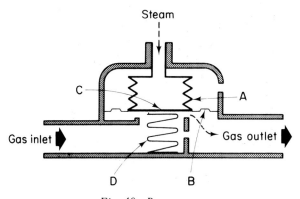

Fig. 49   Pressurestat

the valve against the force of the spring C and allow water to pass to
the ball valve. The spring tension is adjusted to close the valve when
the steam pressure falls below 170 mbar. The valve may be opened
by screwing in the knurled screw at the top in order to fill the boiler

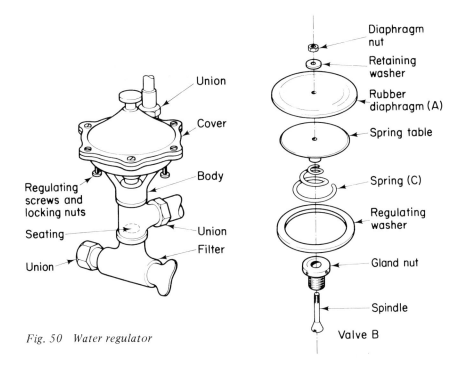

Fig. 50   Water regulator

initially. The screw must then be unscrewed before operating the boiler.

The safety valve, Fig. 51, consists of a valve held against its seating by a spring. Steam pressure on the face of the valve will cause it to open against the tension of the spring and allow steam to be released. The degree of tension determines the relief pressure. On pre-set types the tension nut is screwed down on to a spacing washer of predetermined thickness.

An excess temperature cut-off, similar to that used on some central heating boilers, may be fitted to low water capacity boilers, Fig. 52. It is essentially a heat operated switch which shuts off the main gas and pilot if gas is lit when the boiler is empty.

Gas rates of pressure boilers vary from 8 to 32 kW (27 500 to 109 000 Btu/h).

Café sets produce coffee by percolating it in an infuser, Fig. 53. Freshly ground coffee is placed on a filter paper in the infuser, which is then clamped between the inlet and outlet supplies. Turning the operating cock allows boiling water to flow into the infuser and coffee to flow into the steam jacketed side urn. When the required amount has been made, the cock is turned off, isolating the infuser from the boiler and from the side urn.

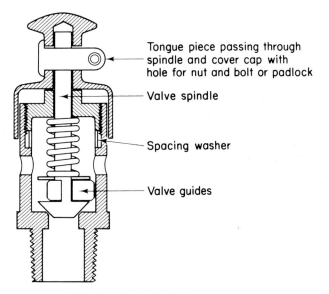

Tongue piece passing through spindle and cover cap with hole for nut and bolt or padlock

Valve spindle

Spacing washer

Valve guides

Fig. 51   Spring-loaded safety valve

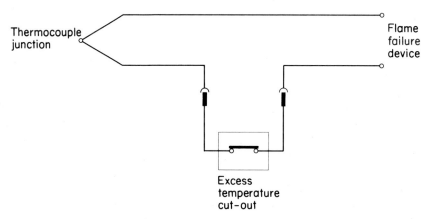

Thermocouple junction

Flame failure device

Excess temperature cut-out

Fig. 52   Excess temperature cut-out

§112 Back Bar Equipment

This is the term used to describe the range of small, counter mounted appliances used in snack bars, cafés and similar small establishments. It embraces griddle plates, fryers, boiling tables, bains-marie and flare grills or salamanders. It could include any appliance required to produce a particular dish.

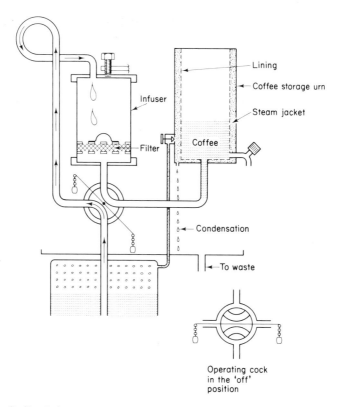

Fig. 53   Coffee infuser

The appliances are miniature versions of those already described, with similar ignition and control devices. Gas rates are smaller in proportion.

§113 Installation of Equipment

General

All installation work must conform to the requirements of the Gas Safety Regulations and the Building Regulations. It should also comply with the recommendations of the relevant British Standard Code of Practice and the appliance manufacturer's instructions.

Appliances should be level, protected from draughts and in a good light. There should be easy access for cleaning and for servicing or adjusting components or controls.

Wall or floor temperatures must not exceed 65°C (149°F) and fire resistant sheets or bases may be necessary with some appliances. Adequate ventilation must be provided (section 114).

In small kitchens the range is the most frequently used appliance and it should be sited in the most convenient position.

Gas Supply

The gas meter and the internal pipework must be adequate to meet the maximum demand. Each appliance should be fitted with a gas control cock so that it may be isolated for servicing. Union cocks are generally provided to allow the appliance to be easily disconnected. Small appliances and back bar equipment may be connected by means of armoured flexible tubing and plug-in connectors to enable them to be withdrawn for cleaning.

Electricity Supply

The electrical installation should comply with the I.E.E. "Regulations for the Electrical Equipment of Buildings".

Supplies for electrical control devices are generally single-phase but single phase motors are usually rated below 1 kW and very rarely up to 5 kW. Many of the motors found on large scale catering equipment are 3-phase.

A three phase supply can be identified from the:

· manufacturer's badge on the appliance
· main switch
· motor starter
· motor connections

Most electrical equipment has a badge stating the type of supply it requires and the voltage at which it operates.

A three phase switch box can usually be identified by three double-bladed contacts, one for each phase, Fig. 54. The box may also incorporate three fuses. Switch boxes can only be opened when the switch is in the off position.

Three phase motors use a "starter" in place of the on-off switch used with single phase. This may be controlled manually or by thermostat, timer or other device. The manual type has two shrouded push-buttons usually towards the top of the box. The left-hand button is black and marked "START", the right-hand button is red and

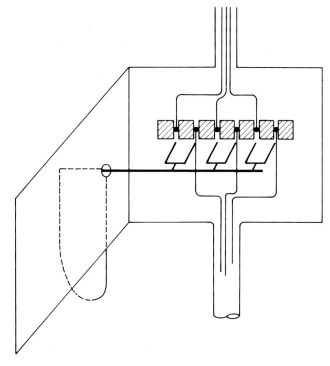

Fig. 54   3-phase switch box

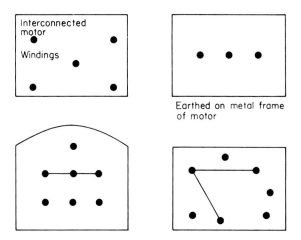

Fig. 55   3-phase motor connections

marked "STOP and RESET". Starters include an overload device which switches the supply off if the motor is overloaded.

Motor connections are not made to any set pattern or colour code.

Typical layouts are shown in Fig. 55. They can be identified by being grouped in threes diagonally, laterally, or in a triangular formation. Resistance readings between the three winding connections should be equal.

The colour code of three phase supplies may be either—

Old Colour Code

The three phases are Red, Blue and Yellow, Neutral is Black.
or—

New Colour Code

The phases are Black or Brown and the terminals are marked L1, L2, L3. Neutral is Blue.

Water Supply

Generally water supplies should be taken from the mains and a stop cock fitted in the line. Where supplies are connected to a cistern, for appliances such as pressure steamers or café boilers, there must be sufficient head of water to operate the appliance.

Waste Pipes

Provision must be made for the removal of waste water and condensation from appliances. Drains may need to incorporate grease traps to prevent blockage.

§114 Ventilation

It is necessary to ensure an adequate supply of fresh air to a kitchen in order to provide air for combustion and to limit the effects of heat and humidity caused by the cooking.

It is also important that products of combustion and cooking vapours are removed at source and not allowed to disperse throughout the kitchen. This is usually achieved by a combination of hoods, ducts and extraction fans. Fresh make-up air must be fed in to take the place of the foul air extracted.

Kitchens require 20 to 40 air changes per hour. Ventilation systems are shown in Fig. 56.

Most catering appliances should be positioned under a hood or canopy which should be sized to extend 200 to 300 mm (8 to 12 in) beyond the equipment. The lower edge of the hood should be about 2 m (6.6 ft) from the floor. It should have a condensation channel around the inside which may be fitted with a drain. Hoods should be made of anodised aluminium, reinforced glass, galvanised

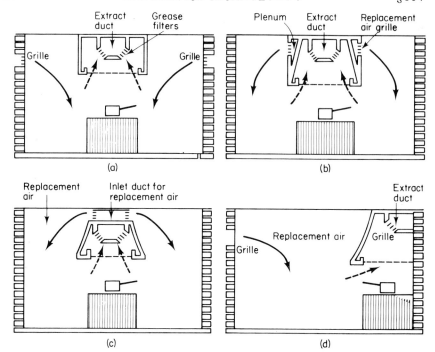

Fig. 56   Ventilating systems (a) traditional extract hood with side wall air inlet
grilles (b) extract hood with air inlet plenum (c) inlet and outlet ducts
in parallel over extract hood (d) side wall hood with cross ventilation of
inlet air

mild steel or stainless steel. The structure must be rigid and stable
and preferably be suspended from the ceiling.

Frying, grilling and roasting produce greasy vapours and cleanable
grease filters should be provided in the flues. They should be cleaned
at least once a month.

All extract ducts should be made of sheet metal and of a gauge
strong enough to withstand frequent scraping and cleaning. Some
grease vapour will pass through the filters and, since its dewpoint is
about 152°C (305°F), will be deposited on the duct walls. The ducts
must have adequate and accessible access doors or panels for cleaning
purposes.

Ducts should be as short and as straight as possible and with an
upward slope of not less than 1 in 12 towards the termination point.

Fans should be mounted on anti-vibration pads and be joined to
the duct by flexible connections in order to minimise noise.

The duct termination outside the building should be selected to
avoid causing any nuisance.

The amount of ventilation air required for various appliances is
given in Table 2.

115   **TABLE 2**   Ventilation Requirements for Catering Appliances

Appliance	Ventilation Rate	
	m³/min	ft³/min
Bain Marie	11	400
Café boiler	14	500
Sterilising sink	14	500
Boiling pan	17	600
Griller—salamander	17	600
Pastry oven	17	600
Range-single oven	17	600
Steaming oven	17	600
Fryer—deep fat	25	900
Grill—underfired	25	900

116   Commissioning

After the installation has been completed, test the gas and water supplies for soundness and purge. Check each appliance in turn for the following points.

Position

Check against any kitchen plans or layout diagrams that the appliance is fitted in the correct location. Ensure that the appliance is level and correctly situated under any extract hoods or over any gulleys or drains.

Assembly

Check against the manufacturer's instructions that:

- appliance is assembled correctly
- doors are properly aligned and door seals effective (re check door seals after heating up)
- correct furniture has been supplied
- mechanical devices operate satisfactorily

Function

Turn on the water supply and fill those appliances which require water. Completely fill café boilers, steamers, bains marie, sterilising sinks and any pan jackets to the required level. Adjust floats or ball valves as necessary.

Boiling pans, bulk water boilers or urns should have sufficient water to cover the base and the draw off connection.

Deep fat fryers should be filled to the correct level with oil. The thermostat shield must be covered.

Turn on the gas supply and light the appliance following the recommended ignition procedure.

Check the following points:

- ignition device operates
- pilot flame stable, correct size, correctly positioned relative to flame failure device
- flame failure device operates correctly
- relight the appliance and check the burner pressure, adjust if necessary
- check flame picture and, if necessary, check gas rate by meter test dial.
- check appliance taps and control devices
- thermostats and relay valves shut down, bypass rates correct
- pressurestats operate at desired pressures
- overheat cut offs and thermal switches are satisfactory
- water regulators maintain correct steam pressure

Check the ventilation system, air inlets clear and adequate, extract fans running smoothly at correct speed and correct direction of rotation. All filters in position and access doors to ducts properly sealed.

Finally, hand over the equipment to the user and ensure that the operating instructions are understood.

§117 Servicing

In most kitchens catering equipment has long periods of heavy use, so regular cleaning and servicing is essential.

Servicing procedures should follow those already established in Volume 2 for comparable domestic appliances. However, because of the degree of use, more attention must be given to those components which are subject to spillage and wear. Items to be checked include the following.

Gas Taps

Check for smooth operation. If necessary, clean out internal gas ways and re-grease.

Draw Off and Drain Taps

Dismantle, clean and re-grease the plug. Check that the connecting pipes are clear.

Float Valves and Ball Valves

Examine moving parts. Check washer and seating for wear and renew or re-seat if necessary. Check operation, confirm water level and adjust if necessary.

Sight Glasses

Dismantle and clean gauge glass tube. Reassemble and renew packing glands. Check water level.

Safety Valves

These may be padlocked and are often ignored. If left for too long they may become scaled up and inoperative. Periodically they should be dismantled, cleaned and have the valve seating ground in. Check that the valve is set to the correct pressure.

Control Devices

Check operation and ensure that gas is cut off at the correct temperature or pressure. Check that bypasses, weep jets and orifices are clear. Renew diaphragms as necessary and repack stuffing glands.

Burners

Clean burners thoroughly. If spillage has carbonised in the burner parts it may be necessary to use a twist drill of the correct diameter or to soak the burner in a caustic solution. After soaking, rinse and dry thoroughly.

Injectors and jets should be cleaned with a brush. A choked orifice may be cleared by using a piece of soft fuse wire.

Flueways

Both internal and external flueways must be kept clear. Internal sections in particular can become encrusted with carbonised grease and must be scraped out. Check that flues are functioning and that flame combustion is satisfactory.

Door Seals

Test door seals and adjust catches or hinges as necessary. Renew worn door gaskets.

Descaling

In hard water areas and if no water softener is used, water boilers will require periodic descaling. Chemical descaling is normally carried out after removing the appliance to a workshop, but most boilers incorporate access facilities for mechanical descaling.

Soft scale can be removed by wire brushing and washing away by water. Hard deposits will require scraping or chipping which must be done carefully with a blunt tool to avoid causing damage.

Manufacturers often offer a descaling service for their appliances.

Ventilation

Check that ventilating system is operating correctly. Clean filters if necessary.

§118   Fault Diagnosis and Remedy

Most of the faults which occur on catering equipment have already been dealt with in the two previous volumes. Additional points which should be noted are listed in the table opposite.

Electrical Faults

Faults on single phase equipment were discussed in Volume 2, a three phase supply presents other problems. A limited amount of testing may be carried out provided that a BGC multi meter with High Voltage Adapter is used.

To test a three phase motor and its supply, check that the motor is free to turn, then carry out the following procedure.

Set the multi meter on the 0 to 300V a.c. scale. Fit the high voltage adaptor and check that the meter operates satisfactorily by testing from L to N on a known 240 V supply, reading the 0 to 6.0 scale.

Isolate the equipment, remove any fuses and hang a notice on the switch box reading "Danger *do not use*. Work in Progress".

Carry out "Phase Voltage Checks" as follows:-

1. Phase to Phase   Red to Yellow (L1 to L2) Red to Blue (L1 to L3) Yellow to Blue (L2 to L3)
2. Phase to Neutral   Red (L1) to N. Blue (L3) to N, Yellow (L2) to N
3. Neutral to Earth   Black (Blue) to E

All readings should be 0 (zero).

If any reading is shown, do not touch the equipment and imme-

Symptom	Action
DEEP FAT FRYERS	
Pilot lights up but goes out when press-button is released	*Check:* • *thermocouple correctly connected* • *overheat thermostat connection satisfactory* • *overheat thermostat correctly calibrated and not faulty*
All gas shuts off during normal working	*Check oil level and top up if necessary* *Check overheat thermostat correctly set and not faulty*
PRESSURE TYPE CAFE BOILER	
Water level too high	*Check float valve:* • *float watertight and not coated with scale* • *valve seating and jumper clean, renew washer is required*
Water level too low	*Check water supply:* • *filter clear* • *regulator set at correct steam pressure and admitting water*
Steam only obtainable from water draw off	*Check water level, check that regulator is admitting water* *Check syphon tube not damaged or missing*
Safety valve blows	*Check pressurestat setting correct* *Check safety valve clean and seating properly, valve set to correct pressure*

diately consult an electrician. There is an electrical fault which must be remedied before any work can be carried out.

If all readings are 0, carry out "Earth Continuity Check", rectify any faults. Check that connections at terminal block and starter/contactor are sound. Check appliance fuses for continuity.

Disconnect output terminal of starter/contactor, set multimeter to $\Omega \times 1$ scale and test from phase to phase, Red to Yellow (L1 to L2), Red to Blue (L1 to L3), Yellow to Blue (L2 to L3).

All readings should be similar.

If one or more reads 0, there is an open circuit.

If one or more reads infinity, there is a short circuit.

In both cases the motor is faulty.

If all readings are similar, set the multimeter to $\Omega \times 100$ scale and test from each phase to earth,

Red (L1) to Earth, Blue (L3) to Earth, Yellow (L2) to Earth

All readings should be infinity

If one or more readings are lower than infinity, the motor is faulty

If all readings are infinity, replace the appliance fuses and re-establish the supply.

Carry out "Phase Voltage Checks" at the appliance terminal block, readings should be :-

Phase to Phase　　•　approximately 415 V a.c.

Phase to Neutral　•　approximately 240 V a.c.

Neutral to Earth　•　approximately　15 V a.c.

If any readings are incorrect there is an electrical fault which must be remedied before continuing.

If readings are correct check that the motor is rotating in the right direction.

If the motor rotates backwards, one phase is reversed

Isolate the equipment as before and reverse any two phase connections, for example, Red (L1) and Blue (L3).

The motor will now work satisfactorily.

Faults on starter/contactors should be referred to an electrician.

§119　Kitchen Planning

The first step in designing the layout of a catering establishment is to allocate areas for the cooking activities and for dining. Calculations are based on the requirements which include :

- kitchen—types of food to be cooked
    —storage facilities required
    —type of service provided
    —number and size of appliances
    —staff facilities, toilets
- dining room—standard of catering
    —size of tables
    —number of diners

Provisional estimates can be based on the figures in Table 3.

120 **TABLE 3** **Approximate Kitchen and Dining Areas**

Kitchen areas include storage and staff accommodation.
Dining areas include gangways.

Number of Persons Catered for	Kitchen Area per Person	
	m²	ft²
100	0.5 to 0.75	5.4 to 8
400	0.3 to 0.5	3.2 to 5.8
1000 and above	0.25 to 0.4	2.7 to 4.3
Number of Persons per Table	Dining Room Area per Person	
	m²	ft²
6 to 8	0.75 to 0.9	8 to 10
4	0.9 to 1	10 to 12

The areas allocated must conform to legislation regarding working space, toilet facilities, fire regulations and other safety measures. Kitchen areas are usually about 30% of the dining area.

121 **Kitchen Layout**

The kitchen should be designed to enable the work activities to proceed in the proper sequence and as smoothly as possible. The work flow should follow the following pattern:

- reception
- storage
- preparation
- cooking
- serving
- washing up

A plan of a typical kitchen is shown in Fig. 57 and a key to the symbols used in Fig. 58.

Reception

Some area is required where incoming goods can be weighted, checked, counted and examined. In small establishments a table may be sufficient.

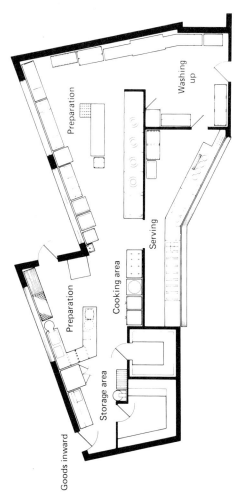

Fig. 57 Typical kitchen layout

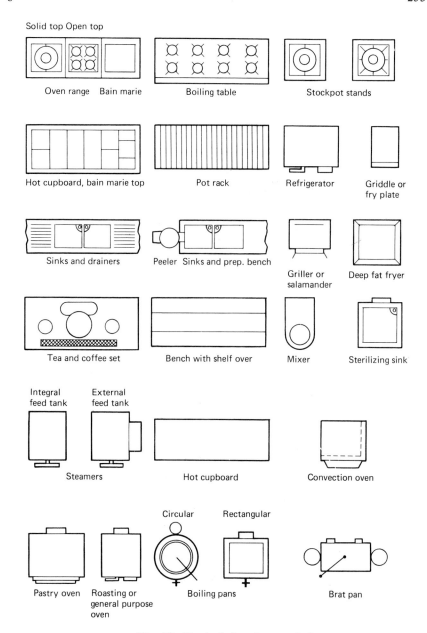

Fig. 58 Typical planning symbols

Storage

It is usual to keep the different kinds of food apart and in large establishments separate stores are required for dry goods, vegetables, meat and fish, frozen foods. Stores should be close to the reception areas.

Dry goods consist of tea, coffee, sugar, flour, tins, packets or jars. Approximate storage areas required are:

- 100 meals/day–0.06 m^2/person (0.6 ft^2/person)
- 1000 meals/day–0.023 m^2/person (0.25 ft^2/person)

Vegetables are stored on racks or duck boards.
Approximate storage required:

- 100 meals/day–0.03 m^2/person (0.3 ft^2/person)
- 1000 meals/day–0.08 m^2/person (0.8 ft^2/person)

Refrigerated storage is required for meat, fish, poultry, dairy produce and cooked food. Approximate refrigeration capacity is from 0.14 to 0.28 m^3/person (0.5 to 1 ft^3/person).

Separate freezer capacity is required for ice cream and frozen food. The average weight which can be stored is $480 \, kg/m^3$ $(30 \, lb/ft^3)$.

Preparation

Preparation areas should be situated between the particular stores and the appropriate cooking appliances. Large kitchens should have separate areas for vegetables, meat and fish, pastry. Preparation areas require work benches, sinks, waste disposal and equipment for cutting, peeling, sawing, mincing and mixing.

Cooking

The conversion of the prepared new food is carried out in separate areas as close as possible to the servery.

Serving

Cooked food is temporarily held in hot cupboards and bains-marie. Depending on the type of service, the servery may also supply hot and cold drinks, cold dishes and ice cream. It may also dispense cutlery and take cash.

Washing Up

This should be situated away from the dining area to reduce noise and be equipped with sinks and machines to clean and sterilise the crockery and cutlery. It needs racks and waste disposal facilities. Clean tableware should be stored adjacent to the servery.

Pot washing is carried out separately and large sinks and racks are required. The clean pots should be stored close to the cooking area.

122   Capacities of Equipment

Average Portions

These may vary considerably between establishments. Some approximate figures are as follows:

Soup or stew	100 portions = 18 to 28 litres	(16 to 25 portions/gallon)
Custard or gravy	100 portions = 6 to 13 litres	(35 to 75 portions/gallon)
Milk pudding	100 portions = 18 litres	(25 portions/gallon)
Potatoes	100 portions = 15 to 18 kg	($2\frac{1}{2}$ to 3 portions/lb)
Green vegetables	100 portions = 15 kg	(3 portions/lb)
Root vegetables	100 portions = 7.5 to 10 kg	($4\frac{1}{2}$ to 6 portions/lb)
Fish	100 portions = 11 to 15 kg	(3 to 4 portions/lb)
Roast meat	100 portions = 9 to 11 kg	(4 to 5 portions/lb)

Chicken or Duck provides 4 to 6 portions
Turkey provides about 30 portions

Beverages

Boiling water required—hotels = 0.28 litres/person
                 ($\frac{1}{2}$ pint/person)
            —canteens = 0.42 litres/person
                 ($\frac{3}{4}$ pint/person)

Tea   100 cups = 19 to 23 litres (20 to 24 cups/gallon)

Coffee 100 cups = 11 litres   (40 cups/gallon)

(milk added extra)

Equipment Sizes and Capacities

1. *Ranges*

         Solid top about 0.84 m² (9 ft²)
         Oven-capacity 0.17 m³ (6 ft³)

    —shelf area 0.31 to 0.46 m² (3 to 5 ft²)

2. *Boiling Tables*

         150 litres/m² (3 gall/ft²)

using vessels of up to 27 litres (6 gall) capacity
Potatoes and root vegetables require 1 litre capacity for every 0.4 kg (1 gallon/4 lb)
Green vegetables require 1 litre for every 0.15 kg (1 gallon/$1\frac{1}{2}$ lb)

3. *Ovens*

   General Purpose or Roasting

   Poultry, 96 to 128 kg/m³ (6 to 8 lb/ft³)

   Roast meat, 128 to 160 kg/m³ (8 to 10 lb/ft³)

   Pies 19.5 kg/m² (4 lb/ft²) shelf area

   or   172 portions/m² (16 portions/ft²)

   Pastry

   Small cakes, 250/m² (25/ft²)

   Meat pies, 108 × 0.11 kg pies/m²
   (10 × 4 oz pies/ft²)

4. *Grills*

   Underfired

   Steaks, 3 to 6 minutes, 65 to 85/m² (6 to 8/ft²)

   Overfired

   Steaks, 2 to 4 minutes with branding plate capacity as the underfired type.

   Toast, $1\frac{1}{2}$ to 2 minutes, 65 slices/m² (6 slices/ft²)

6. *Fryers*

   Deep Fat

   Fish or chips 14.6 kg/m² (3 lb/ft²)

   Brat Pans

   Dry fry, eggs,   600/hour

   bacon, 600 portions/hour

   Poach, eggs,   800/hour

   fish,   200 portions/hour

   Sauté   meat,   27 kg/hour (60 lb/hour)

7. *Steamers*

   Puddings,
   individual, 172 × 75 mm/m² (16 × 3 in/ft²) shelf area
   roll tins, 258 portions/m² (24 portions/ft²) shelf area

Root vegetables,   384 kg/m³ (24 lb/ft³)
Fish,   72 kg/m³ (4.5 lb/ft³)

8. *Boiling Pans*

Green vegetables, 0.15 kg/litre ($1\frac{1}{2}$ lb/gallon)

Root vegetables, 0.4 kg/litre (4 lb/gallon)

Equipment Requirements

One single oven range will cater for up to 50 persons. A double oven range is normally adequate for 100 persons although a separate over-fired grill might be a useful addition. Above 100 persons more specialised equipment should be used, the type of appliance varying with the menu to be offered.

123   Cook/Freeze Catering

"Cook/freeze" is the term used to describe frozen food production. Food is cooked, frozen quickly and put into cold storage. It may then be reheated and served at a later date and often at another location.

The sequence of operations is:-

- receive
- store
- prepare
- cook
- portion
- pack
- freeze
- store

The cooking appliances used may be the conventional types and may include:

- large boiling pans (135 to 180 litres (30 to 40 gall))
- steamers
- convection ovens
- deep fat fryers
- baking ovens

Large quantities of food are cooked at a time, possibly 500 portions. For very large quantities factory methods are applied and food travels on conveyors, through frying oil, boiling water or ovens as appropriate.

After cooking, the food is divided into individual portions and sealed into plastic pouches or aluminium foil containers. The containers are labelled and coded before freezing. All precautions are taken to avoid hygiene hazards.

Freezing is usually carried out in a "blast freezer". This is an insulated tunnel in which refrigerated air is blown over the food at velocities of about 5 m/second or 1000 ft/min. Trolley loads of food have their temperature reduced to $-20°C$ ($-4°F$) in about 90 minutes. An alternative method used particularly for soft fruits, is to spray the food with liquid nitrogen at $-196°C$ ($-321°F$). This produces almost instantaneous freezing. Frozen food is stored at about $-20°C$ ($-4°F$).

The "end" or "finishing" kitchens complete the cycle. They require a cold store and usually use forced convection ovens for reheating the frozen food. They also need ancillary equipment for cooking frozen vegetables, frozen chips and fried meals. In addition, they have to provide soup, sweets and beverages and will require a servery.

Cook/freeze systems may vary depending on the type and quantity of the foods produced. This determines the amount and type of equipment required by individual caterers.

CHAPTER 7

Incinerators

Chapter 7 is based on an original draft by Mr. R. L. Brooks

---

124 Introduction

Incineration is the process of burning, so as to reduce to ashes. An "incinerator" is a specially constructed furnace for carrying out this process in order to dispose of waste materials. The residue is sterile, odourless and reduced to between 7 and 12% of its original volume. It contains little or no combustible material.

Gas is often used as a fuel for incineration and there are many different types of incinerators available to deal with the various forms of waste materials. These materials may be classified in a number of ways. A method adopted by the British Standards Institution makes use of three of the six American definitions. These are for rubbish, refuse and garbage. "Refuse" represents household waste and contains equal parts of rubbish and garbage.

Table 1 gives details of the classification.

This classification can be simplified by dividing waste into two main categories:

- harmless, inoffensive waste which ignites easily and burns freely (rubbish)
- organic wastes which may be offensive, do not ignite easily and require additional fuel for combustion (garbage)

Most other common types of waste are simply a combination of these two, in varying proportions. Industrial wastes may be any of a wide range of substances which could be obnoxious or toxic. Specially designed plant may be necessary for their disposal. All types of waste can create hazards. Rubbish can present a fire risk and garbage a rise of infection. Incineration can eliminate these risks.

126 Types of Incinerator

Incinerators are designed for a variety of applications including:

- sanitary
- domestic; individual or multi-storey dwellings

§125 **TABLE 1** Classification of Waste Materials for Incineration

Class	Type	Composition Approximate % by Weight	Constituents	Source	Moisture Content %	Calorific Value MJ/kg	Fuel Required to Incinerate kW/kg	Incombustible Solids % Volume
1	Rubbish	Rubbish up to 100% Garbage up to 20%	Combustible waste, paper, cartons, wood scraps, rags, floor sweepings	Domestic, commercial or industrial	25%	15.15	0	10%
2	Refuse	Rubbish 50% Garbage 50%	Rubbish and garbage	Residential	50%	10.02	0	7%
3	Garbage	Rubbish up to 35% Garbage up to 100%	Animal and vegetable wastes	Restaurants, hotels, clubs, institutions markets, commercial premises	70%	2.8	0.64	5%

- hospital; ward or general
- large-scale commercial, industrial or municipal
- cremation

Gas fired incinerators are generally classified as commercial appliances and all sizes have commercial applications. They may be divided into four main classes, depending on their capacity and the type of waste incinerated.

These are:

- sanitary
- domestic
- general
- specialised

Sanitary (Fig. 1)

Capacity,   0.003 to 0.009 m³ (0.1 to 0.3 ft³)
Type of waste,   sanitary towels and other small dressings
Heat input,   1.75 to 5.85 kW (6000 to 20 000 Btu/h)
Usage,   home, offices, factories, hospitals, institutions, public or communal toilets.

Fig. 1 Sanitary incinerators

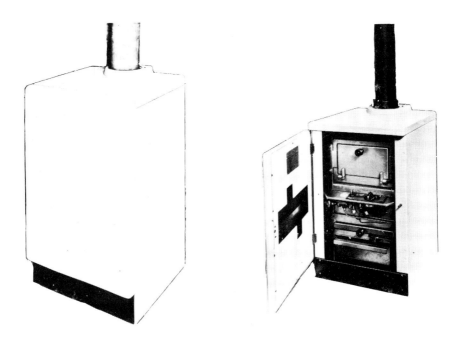

Fig. 2   *Domestic incinerator, 0.02 m³*

Fig. 2(a)   Domestic incinerator, 0.06 m³

Domestic, (Fig. 2 and 2(a))

Capacity,	0.02 to 0.06 m^3 (0.7 to 2 ft^3)
Type of waste,	household refuse, kitchen waste
Heat input,	5.85 to 10 kW (20 000 to 34 000 Btu/h)
Usage,	homes, the larger sizes may be used communally in blocks or flats, or in hospital wards

General (Fig. 3)

Capacity,	0.07 m^3 and upwards (2.5 ft^3 upwards)
Type of waste,	hospitals, and general commercial wastes
Heat input,	15 kW upwards (51 000 Btu/h upwards)
Usage,	hospitals, stores, supermarkets, cinemas, schools

Fig. 3   General incinerator

Specialised

These vary in capacity and design. They may be capable of handling up to 1000 kg (1 ton) of waste per hour. Special features can include:

- secondary burners to consume smoke and fumes
- fanned draught flues
- automatic packaged burners
- full sequence controls

- timed operation
- water sprays and sump to remove fly ash from the flue
- facilities for pre-drying waste before ignition
- automatic adjustment of heat input
- control of flue temperatures
- utilisation of waste heat from the process

Cremators are a specialised form of incinerator.

They may be designed for pathological or veterinary wastes as well as for human cremations. They are similar to open hearth furnaces and operate at high temperatures. Towards the end of the process the gas may be shut off and air alone fed in to complete the combustion.

§127   Principle of Operation

Figure 4 illustrates the essential components and the operation of a medium sized gas incinerator. It consists of a primary chamber with a loading door near the top and grate bars at the bottom on to which the waste is placed.

In the type shown, a primary burner for drying and igniting the load is positioned below the bars and fires through them. As the load burns, ash falls between the grate bars into an ash pan or box below.

Smoke and fumes from the primary chamber are drawn, by the

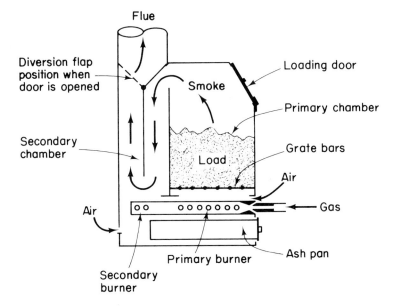

Fig. 4   Operation of incinerator

flue "pull', down into the secondary chamber. Here the smoke and odours are consumed by the flame from a secondary burner or "after burner". This reduces the smoke to superheated steam and carbon dioxide. It also helps to prevent condensation.

Finally, the odourless and smokeless waste gases are evacuated by the flue.

128 Construction of Incinerators

Incinerators of certain types are designed to comply with the requirements of British Standard Specifications.
These include:

- BS 3107 "Small incinerators for the destruction of hospital dressings"
- BS 3316 "Large incinerators for the destruction of hospital waste"
- BS 3813 "Incinerators for waste from trade and residential premises. Capacities from 22.7 to 453.6 kg/h (50 to 1000 lb/h)"

All three standards include requirements for:

- surface and door handle maximum temperatures
- size and position of charging opening
- arrangement of pilot and flame protection device

Other requirements are specific to the class of incinerator as follows:

- BS 3107 —height and thickness of ash pan
        —minimum size of flue
        —performance and method of test
- BS 3316 —secondary chamber burner
        —dust arrestor
- BS 3813 —smoke, dust and grit emission to comply with the Clean Air Act

The construction of the different types of incinerator vary quite considerably and are, therefore, dealt with under separate headings.

129 Sanitary Incinerators

Construction

The general construction of a small sanitary incinerator is shown in Fig. 5. The primary chamber is of moulded or fabricated refractory brick and is mounted above heat resistant cast iron or steel grate

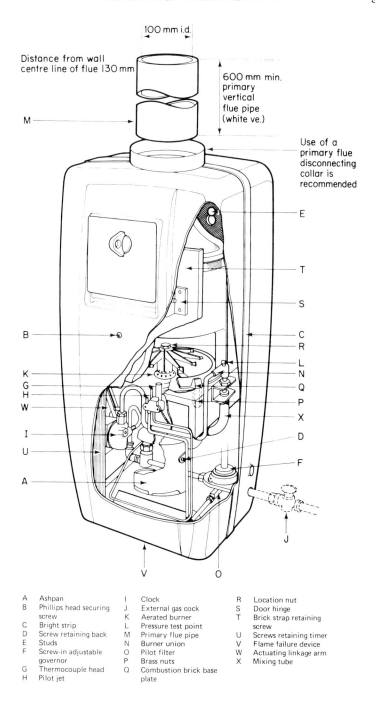

100 mm i.d.

Distance from wall
centre line of flue 130 mm

600 mm min.
primary
vertical
flue pipe
(white ve.)

M

Use of a
primary flue
disconnecting
collar is
recommended

E

T

S

B

C
R

K

L
N

G
H

Q

W

P

I

X

U

D

A

F

J

V   O

A	Ashpan	I	Clock	R	Location nut
B	Phillips head securing screw	J	External gas cock	S	Door hinge
C	Bright strip	K	Aerated burner	T	Brick strap retaining screw
D	Screw retaining back	L	Pressure test point	U	Screws retaining timer
E	Studs	M	Primary flue pipe	V	Flame failure device
F	Screw-in adjustable governor	N	Burner union	W	Actuating linkage arm
G	Thermocouple head	O	Pilot filter	X	Mixing tube
H	Pilot jet	P	Brass nuts		
		Q	Combustion brick base plate		

Fig. 5 Construction of small sanitary incinerator

bars. The burner is a natural draught drilled ring which is positioned centrally below the grate bars.

Waste is fed in through a front, drop down loading door. This may be fitted with a hopper to limit the size of the load and keep the primary chamber closed while the door is open. The door is linked mechanically to the timer and to the grate bars. Opening and closing the door starts the timer, so turning on the gas to the primary burner. This then fires for about 15 to 20 minutes. Movement of the loading door also shakes the grate bars to break up the ashes so that they fall into the ash pan below. The ash pan may be removable either through a front door or, as illustrated, from underneath. The capacity of the pan is required by BS 3107 to hold the volume of ash resulting from one week of use.

Small sanitary incinerators generally do not have a secondary chamber or after burner. This is because the incineration of small dressings is mainly the evaporation of a little moisture with only a small weight of solid matter to be finally burnt. Combustion is normally complete within the primary chamber and very little smoke is produced.

The outer case is of steel sheet usually stove enamelled. The incinerator is wall fixing, suspended from slots in the near section of the case.

The gas supply connection is $R_c \frac{1}{4}$ into a governor and a control cock must be fitted. The flue is 100 mm (4 in) diameter. The heater is suitable for natural or mechanical draught from 0.57 to 2 m³/min (20 to 70 ft³/min).

Controls and Operation

The controls are located in the lower section of the appliance and include:

- governor
- timer
- thermo electric flame protection device
- solenoid valve and air flow switch, if flued by mechanical extraction

The original timers were air dashpots with an adjustable needle valve to control the timing as in Fig. 6. Later models used clockwork timers operating a gas valve. More recently, clockwork timers are used which operate an electric switch which interrupts the thermocouple circuit.

Ignition is usually manual to a permanent pilot. With the pilot established and the flame protection valve operating, opening the loading door will riddle the grate, wind the timer spring (or raise the

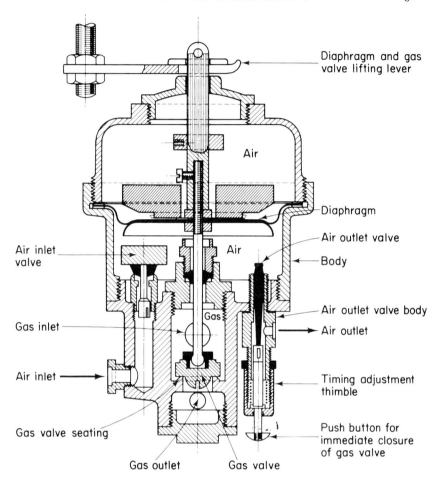

Fig. 6   Air dashpot timing device

diaphragm of a dashpot timer) and open the main gas supply to the burner. The loading door must be opened fully or the timer will not be properly set. This will result in a shorter firing period and probably incomplete incineration of the load.

Closing the door allows the clock spring to drive the timer escapement (or the diaphragm of the dashpot to fall) so that gas to the burner will be cut off after a pre-set time determined by the clock setting or the adjustment of the dashpot air valve.

Disposal Rate

Small incinerators of about $0.009 \, m^3$ $(0.3 \, ft^3)$ can dispose of about 30 single dressings per hour. A bulk load of 16 dressings should be disposed of in about 15 minutes.

It is usual to allow one small sanitary incinerator for every 25 female staff.

130 Domestic Incinerators

Construction

A small domestic incinerator is pictured in Fig. 7. The outer case has been removed to show the construction and the controls.

The primary chamber is lined with firebrick and has a front, drop down loading door. This may be inter-locked with the timer and also operate a flue flap or "diversion" flap. Opening the door moves the

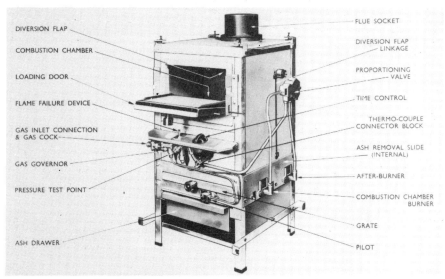

Fig. 7 Construction of small domestic incinerator

flap to connect the primary chamber directly to the flue and so prevent flue gases escaping through the open door. Fig. 4 shows the operation of the flap.

The cast iron grate is a single component and may be riddled by moving the front knob backwards and forwards. Beneath the solid section of the grate is a duplex drilled bar burner, the end portion of which projects into the secondary chamber. Some models use a burner with a single injector as in Fig. 4. Where drilled bar burners are mounted below grate bars, they may incorporate a flange to protect the parts from falling ash, Fig. 8.

The secondary chamber is of heat resistant steel and is located behind the primary chamber. The flue gases pass down to the secondary burner and undergo a change of direction of 180° as they turn

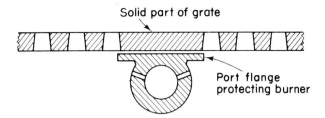

Fig. 8   Drilled bar burner beneath grate

upwards to the flue. This allows any dust or fly ash carried in the gas stream to be thrown to the bottom of the secondary chamber, so it reduces the emission of solid material from the terminal.

The need for after burners and a change of flow direction to clean the flue gases of domestic incinerators was established by research at the Battelle Memorial Institute in America during the mid-fifties. Since then both American and British designs have incorporated these features.

The outer case is usually of sheet steel, stove enamelled. Some larger models may be vitreous enamelled. Small incinerators may be wall or floor fixing. Larger types are freestanding.

The gas connection is Rc $\frac{1}{2}$ into a control cock and the flue is 100 mm (4 in) diameter, heavy duty asbestos cement, or cast iron. Fanned draught may be used.

Controls and Operation

The controls are mainly located at the front of the appliance and are accessible on opening the front door. The side panel must be removed to expose the burner and the proportioning valve.

The controls include:

* gas control cock
* governor
* timer
* thermo electric flame protection device
* bi-metal proportioning valve
* solenoid valve and air flow switch, if flued by fanned draught

Timers are generally clockwork and provide for burning periods of up to four to six hours. The longer times are needed for the disposal of wet kitchen waste. On the larger models the timer may control a weep line from a relay valve.

The proportioning valve is an overriding control which reduces the gas rate as combustion proceeds. When most of the water has been evaporated and the load begins to burn, the temperature rises rapidly.

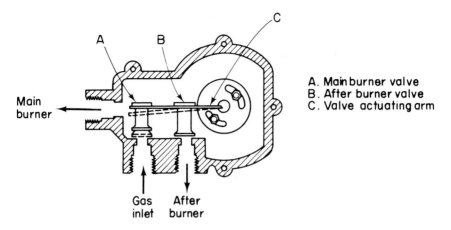

A. Main burner valve
B. After burner valve
C. Valve actuating arm

Fig. 9 Proportioning valve

Less smoke is produced and eventually the after burner is no longer required. The proportioning valve is essentially a bi-metal thermostat which prevents the grate or the flue gases from becoming overheated. It also ensures economy in the use of gas.

The valve is shown in Fig. 9. It consists of a bimetal spiral, set in a sheath and mounted in the secondary chamber flueway. Movement of the spiral rotates a spindle and moves an arm on which are mounted two valves. One valve controls the gas inlet to both burners and the other, the outlet to the after burner. Both valves are bypassed and are spring loaded to allow the arm to overrun if the temperature becomes excessive. The valves are positioned so that the after burner is turned off first, followed later by the main burner. The bypasses ensure that the flames are not completely extinguished until turned off by the timer.

With the timer set to zero the pilot can be ignited, usually via the ash drawer compartment. Opening the loading door allows the waste to be loaded and when the door is closed, a burning time suitable for the load is selected either by a sliding or rotary knob. Finally the starting button is pressed. This opens the valve on the flame protection device and, if the pilot has heated the thermocouple, the valve will remain open until the timer breaks the thermocouple circuit. The valve then closes, leaving only the permanent pilot alight.

Disposal rate

An incinerator of this type and 0.06 m³ (2 ft³) capacity, used for the bulk disposal of sanitary dressings, would deal with 13 kg (26 lb) over 8 hours.

§131 General Incinerators

Construction

A small general incinerator is shown in Fig. 10. This is similar to the larger domestic incinerators and has a single-port, natural draught burner firing through a tunnel which has flame ports in the primary chamber. The end of the tunnel enters the secondary chamber so that one flame serves as primary and secondary burner. The single burner may be sited below, on, or above the grate bars as in Fig. 11.

This size of incinerator may have a cast iron primary chamber as in Fig. 10 but larger models use refractories for both primary and secondary chambers.

Top loading doors are commonly used in smaller models and may be locked by the timing device until the burning period has elapsed.

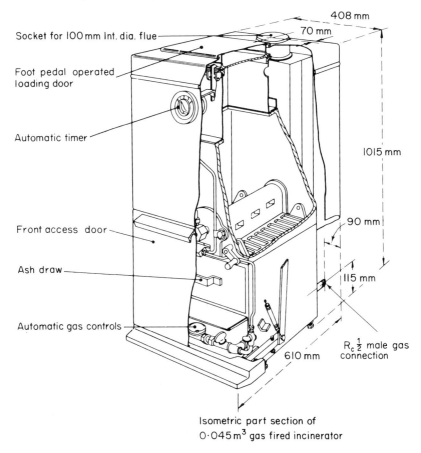

Isometric part section of
$0.045\,m^3$ gas fired incinerator

Fig. 10   Construction of small general incinerator

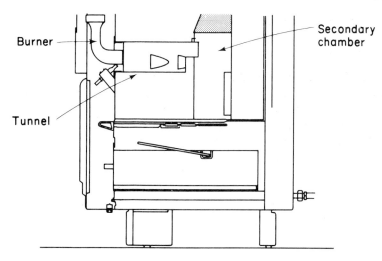

Fig. 11   Single Port burner in tunnel

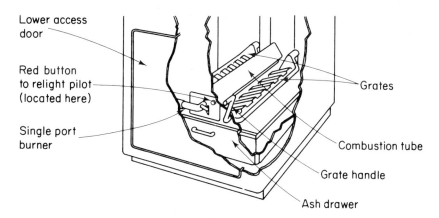

Fig. 12   Rocking grate bars

The grate bars are moveable and may be of the rocking type, Fig. 12.

An incinerator with a capacity of $0.14\,\text{m}^3$ ($5\,\text{ft}^3$) is shown in Fig. 13. This has a single port primary burner firing below the grate bars and a secondary burner firing into a refractory secondary chamber. To economise in floor space the secondary chamber is above the primary chamber. This compact arrangement has been used in sizes up to about $0.5\,\text{m}^3$ ($17\,\text{ft}^3$) capacity. The separate connections to the burners are $R_c 1$ and the flue is $150\,\text{mm}$ ($6\,\text{in}$) cast iron.

The smaller types use natural draught burners but above $0.4\,\text{m}^3$

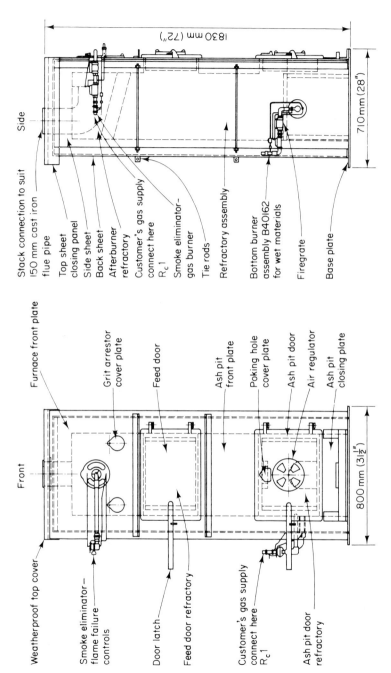

Side

Stack connection to suit
150 mm cast iron
flue pipe

Top sheet
closing panel

Side sheet

Back sheet

Afterburner
refractory

Customer's gas supply
connect here
$R_c 1$

Smoke eliminator–
gas burner

Tie rods

Refractory assembly

Bottom burner
assembly B40162
for wet materials

Firegrate

Base plate

1830 mm (72")

710mm (28")

Furnace front plate

Grit arrestor
cover plate

Feed door

Ash pit
front plate

Poking hole
cover plate

Ash pit door

Air regulator

Ash pit
closing plate

Front

Weatherproof top cover

Smoke eliminator–
flame failure
controls

Door latch

Feed door refractory

Customer's gas supply
connect here
$R_c 1$

Ash pit door
refractory

800 mm ($3\frac{1}{2}$")

Fig. 13   General incinerator, 014 m³

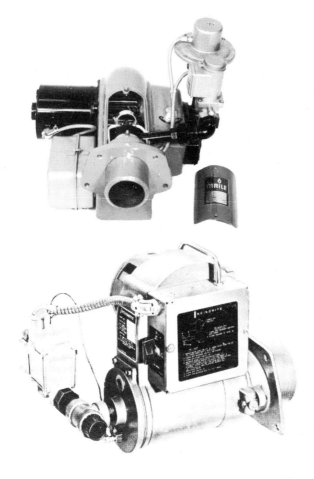

Fig. 14   Forced draught burners

(15 ft³) capacity fanned draught packaged burners are fitted. Two typical burners are shown in Fig. 14.

The range of sizes up to 1.7 m³ (60 ft³) includes the largest hospital incinerators as shown in Fig. 15. These have the more usual arrangement with a secondary chamber behind the primary chamber. Both chambers are lined with refractory brickwork.

The secondary chamber may incorporate additional air ports as well as the secondary burner. The 180° turn in the flow of the flue gases takes place above a water trough. The fly ash and any solid particles thrown out are trapped in the water, forming a sludge which is easily removed.

Some models dispense with the trough and use a dry cyclone to

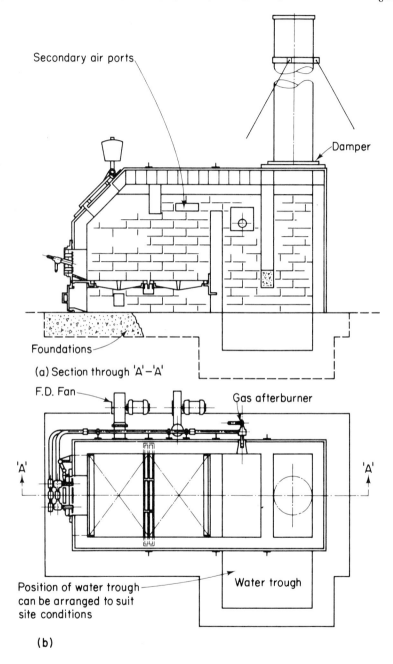

Secondary air ports

Damper

Foundations

(a) Section through 'A'–'A'

F.D. Fan

Gas afterburner

'A'   'A'

Position of water trough
can be arranged to suit
site conditions

Water trough

(b)

Fig. 15   Large hospital incinerator (a) side section (b) plan section

remove the dust burden, particularly when a fan is used to assist air flow through the plant. Others have additional flue gas washing sprays and settling chambers.

A crematory hearth with extra burners may be added to deal with theatre waste.

Controls and Operation

Natural draught burners usually have thermo-electric flame protection devices with manual ignition to a pilot burner. Fanned draught burners have automatic spark ignition with electronic flame sensing and full sequence controls.

The small general incinerators are operated in a similar way to the domestic models. Large plant which requires a flow of air or water during incineration will have flow switches in those lines, interlocking with the burner controls. Until the flows are established the burner cannot be started up.

Disposal Rate

The incineration of general waste is normally a simple process and may be carried out at a rate depending on the area of the grate bars.

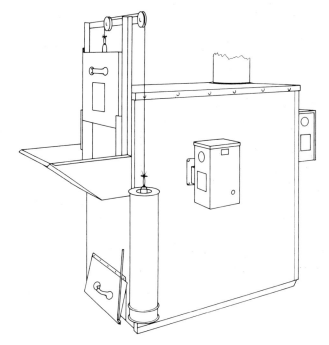

Fig. 16 Hopper feed

The average weight of waste incinerated per hour is:

- natural draught, 122 kg/m² of grate area
  (25 lb/ft²)
- forced draught, 244 kg/m² of grate area
  (50 lb/ft²)

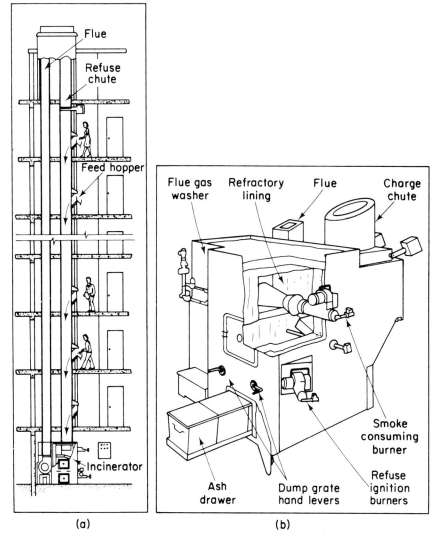

Fig. 17   Refuse chute and incinerator for high-rise flats (a) refuse chute
(b) incinerator

132   Multi-Storey Installations

The on-site disposal of refuse from multi-storey flats may be carried out by a general incinerator similar to that described for hospital waste. The appliance may be hopper fed with a charging door, Fig. 16, or fed from a refuse chute, Fig. 17.

The chute has charging hoppers on each landing which hold the contents of a normal pedal bin and which close the chute opening when the hopper is fully open. Refuse from the chute falls on to a sloping hearth on which wet waste is dried out before falling on to the grate bars. The bars may be tipped to direct the ash into bins in the ash drawer below.

Flue gas washing sprays clean the fly ash from the waste gases and it is collected from a water sump. A flue fan discharges the products up the flue and also maintains a suction on the refuse chute and the primary chamber to prevent odours escaping.

Both primary and secondary chambers are fired by fully automatic packaged burners with protection against failure of the flames or interruption of gas or electrical supplies.

The plant is controlled by a clock which can start the incinerator at about 7 a.m. and shut it off at about 10 p.m. A second clock controls the primary burner, switching it on for short periods at regular intervals during the day.

133   Installing Incinerators

Location

The location of natural draught sanitary, domestic and small general incinerators is usually dictated by the flue run. So the appliance is normally sited on or adjacent to, an outside wall. A natural draught appliance must not be fitted in a room which has an extractor fan.

Where a mechanical extraction flue system is used, mainly in multi-storey buildings, the choice of site is widened.

Small sanitary incinerators are normally installed in communal toilets. Where there is more than one cubicle the appliance may be fitted in the room or in one of the cubicles with an appropriate sign on the door. An air inlet is required to provide combustion air either by natural or mechanical ventilation.

Large general incinerators are usually housed in purpose built locations. The requirements for their installation are similar to those for any gas fired furnace. Because of the large amount of refractory brickwork involved, this type is often constructed on site.

General

Note should be taken of the recommendations and statutory requirements of the following documents:

- the Building Regulations or the Building Standards (Scotland) Regulations
- the Gas Safety Regulations
- IGDC Report No. 718/59 "The installation of small gas fired incinerators"
- current BS codes of practice for flues and installation work
- the Clean Air Act

When incinerators are to be installed the approval of the local authority should be obtained. Generally if a proposed installation meets the requirements of the Building Regulations it is likely to meet those of the local authority.

Flues

Natural draught systems. Incinerator flues are never fitted with draught diverters. This is in order to:

- prevent spillage or products of combustion or particles of burnt rubbish
- ensure that an adequate draught is maintained to give complete incineration

Some incinerators are fitted with a draught stabiliser, Fig. 18. This is a freely pivoted damper which is held closed by a weight. When the flue pull becomes excessive, as when a mainly combustible load is

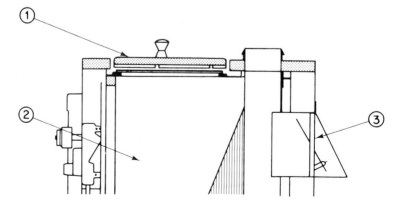

1. Loading door
2. Primary chamber
3. Draught stabiliser

Fig. 18   Draught stabiliser

burnt, the damper opens to allow cool air into the flue. This has the effect of:

- reducing the flue gas temperature
- reducing excessive flue pull
- preventing over fast combustion
- preventing overheating of the appliance

Natural draught flues for incinerators should conform to the general requirements of BS 5440 : Part 1, which are given in Volume 2, Chapter 5. They are only suitable for single appliances, shared flue systems should have mechanical extraction.

The vertical height should be not less than 1.5 m (5 ft) and the minimum diameter 100 mm (4 in). If the total length exceeds 15 m (50 ft) the diameter can be increased to 150 mm (6 in). The larger size reduces the risk of blockage.

The flue should be vertical, as far as possible. No part should be at an angle of less than 45° to the horizontal and only obtuse bends should be used. The amount of external flue should be kept to a minimum.

The Building Regulations require that provision is made for cleaning and inspecting incinerator flues. All bends should be of a type fitted with cleaning door.

Aluminium flue components should not be used. An asbestos cement flue serving an incinerator of a capacity above 0.03 m³ (1 ft³) should be heavy quality and conform to BS 835.

Flue pipe sockets should face the terminal and all joints should be well caulked. Existing brick flues may be used provided that:

- the brickwork is in good condition
- the flue does not communicate with another part of the building
- there are no obstructions
- all other apertures are closed

Mechanical extraction systems. Mechanical extraction may be used either for venting individual appliances or for a number of appliances sharing a common flue.

On an individual appliance where a mechanical system is required or where a natural draught system has failed to operate satisfactorily, mechanical extraction may be used to supplement natural draught. If the flue is designed for natural draught then a small fan may be fitted in the flue duct giving a flow of 0.6 to 2.4 m³/min (20 to 65 ft³/min) against a pressure of 0.5 mbar (0.2 in w.g.).

It is not necessary for any automatic controls to be fitted in this case, a simple on/off switch is adequate.

With multiple appliances sharing a common flue, the main flue must be sized, depending on the number of appliances connected.

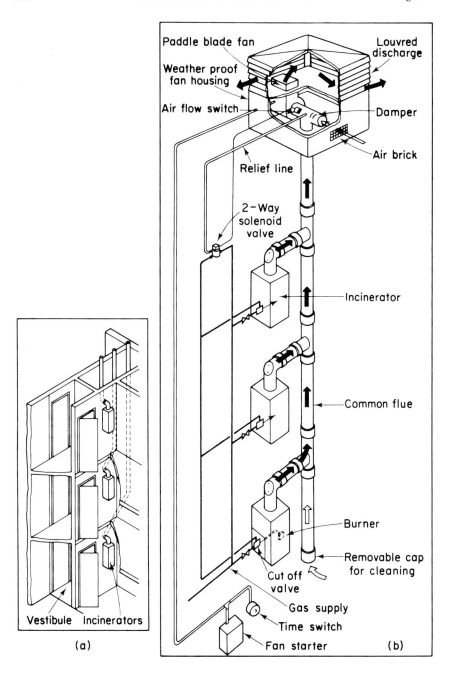

Fig. 19   Flue systems (a) individual flues (b) common flue with mechanical extraction

A better performance may be obtained from a flue of constant diameter, rather than a stepped flue. The flow through each incinerator must be balanced and flue restrictors may need to be fitted to those appliances nearest to the fan. Restrictors of various sizes are available for use with sanitary incinerators. A method of calculating the flow in each limb is given in IGDC Report No. 718/59. Any mechanical ventilation of the room should be taken into consideration when designing an extraction system.

Great care must be taken during the installation to ensure that the flue joints are sound. Leakage into the flue can seriously affect its performance.

When mechanical extraction is used, each appliance must be fitted with a flame protection device and an interlock system to cut off the gas supply to each appliance in the event of fan failure. A system is shown in Fig. 19. This uses the two-way solenoid valve in conjunction with cut off valves on each appliance, Fig. 20. When air pressure fails, the solenoid closes, opening up its lower valve and allowing gas pressure to build up above the cut off diaphragms, so shutting off the appliances.

Alternatively, solenoid valves may be fitted to each appliance, controlled by an air flow switch mounted near to the inlet of the fan.

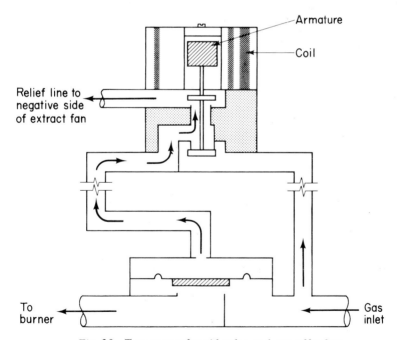

Fig. 20 Two-way solenoid valve and cut-off valves

§134   Commissioning

The procedure for commissioning incinerators will vary with the type and the flue system. Manufacturer's instructions should always be followed. On large installations the commissioning will usually be carried out by the manufacturer or the contractor responsible for building the equipment.

For smaller incinerators, the procedure includes the following operations:

- check the gas supply for soundness and purge
- light the appliance in accordance with the manufacturer's instructions; test any ignition device
- check pilot flame established, stable and correctly positioned relative to flame monitoring sensor
- light up primary and secondary burners in turn, check burner pressure and gas rate at the meter if necessary
  (where a proportioning valve is fitted, any adjustment must be made in the first two minutes after lighting)
- check controls, as fitted
  −flame protection device and flame sensors
  −timer
  −proportioning valve
  −flue thermostat controlling secondary burner
  −safety shut off valves and flow switches on:-
    - mechanical extraction
    - flue gas washer water supply
  −clock controls
- check operation of loading door, diversion flap and adjustment of linkages and interlocks

When the appliance is operational it can be checked on its performance when incinerating an average load of appropriate waste. Check that the load is completely burnt in the recommended time. Check that the flue gases are extracted satisfactorily without leakage or spillage and that no smoke is emitted. Check the operation of any draught stabiliser.

Finally, instruct the operator, set any clock controls and leave the appliance in operation.

§135   Servicing

The servicing of the smaller incinerators is mainly a cleaning operation, particularly where the pilot and main burner are below the grate bars.

With small sanitary incinerators it is often desirable to service the burners and controls away from the appliance. These components can usually be removed as a unit and replaced by a spare. The work on site is reduced to cleaning out the combustion chamber, followed by an inspection of the refractory lining and flueways and attention to the door mechanism and linkages.

The general procedure for servicing domestic and general incinerators should include the following operations:

- question the operator on the performance of the appliance
- remove outer case or panels as required
- riddle the grate, check for freedom of movement, primary chamber clear of ash
- remove and empty ash drawer or pan
- remove and clean pilot burner
- remove and clean primary and secondary burners
- check that injectors are clear, burner tunnel clear and ports free from blockage
- examine condition of refractory lining to primary chamber
- clear any dust from secondary chamber
- check flueways clear
- clean and check door mechanism
- clean linkages and interlocks and adjust if necessary
- check gas control cocks, ease and grease if required
- replace gas burners
- check gas soundness
- light the appliance and check any ignition device
- check pilot stable and correctly positioned
- check burner pressures or gas rate at the meter if necessary
- check flue for pull and spillage
- check controls
    —flame protection device
    —timer set correctly
    —any other control devices
- check that the timer returns to zero, shutting down the appliance at the end of the combustion cycle.

136   Fault Diagnosis and Remedy

Most of the faults which can occur on incinerators are associated with the burners or the control devices and have been dealt with in previous chapters or volumes.

Because of the amount of dust and debris associated with incineration, a number of faults may be due to accumulations of fly ash in burners and flame ports or deposits on jets or thermocouples.

Other common faults are:

Timing Erratic

If the timer is operated by the loading door or by a remote knob, the linkage may be slack and require adjustment.

If the linkage is satisfactory, then the fault is in the timer which should be exchanged. It may be possible to repair an early dash pot type of timer but repairs to clockwork timers should not be attempted.

Smoke Issuing from the Incinerator

This is due to inadequate flue pull. If the flue has previously worked satisfactorily, it is probably due to partial blockage. Check the flue, starting at the incinerator. Blockages most frequently occur near bends.

Excessive Smoke from Terminal

This is usually due to the secondary burner failing to consume the smoke. It is commonly caused by overloading the combustion chamber, a blockage in the burner or an inadequate gas rate. The combustion chamber should be emptied, the burner cleaned and readjusted or the flueway and secondary air supply checked as appropriate.

On small sanitary incinerators, the fault lies with the main gas burner.

The emission of smoke, dust and grit is subject to the requirements of the Clean Air Acts 1956 and 1968.

The principal requirements of the 1956 Act is that dark smoke (40% obscuration Ringleman 2) shall not be emitted from the chimney of any building, including dwellings. It is smoke of this, or greater density which must be destroyed by a secondary burner.

There is some relaxation for industrial plant in the 1968 Act but, unless for soot blowing, it is an offence to emit dark smoke for more than 4 minutes or black smoke (80% obscuration Ringleman 4) for more than a total of 2 minutes in any period of 30 minutes.

The 1968 Act also extends the dark smoke requirements to premises and sites. It makes it an offence not to use all practicable means to reduce the emission of dust and grit from chimneys. Domestic furnaces are exempted if the heating capacity is not more than 16 kW (54 500 Btu/h).

The Act also requires that furnaces burning solid fuel or waste at 50 kg/h (110 lb/h) or more shall be fitted with grit arresting plant, approved by the local authority.

The statutory requirements are frequently revised and the installation and operation of any gas fired incinerator must be carried out in accordance with the current ligislation.

Steam Boilers

Chapter 8 is based on an original draft by Mr. A. J. Spackman

§137 Introduction

Steam is extensively used in industry and commerce chiefly for heating processes or for power generation.

Its main characteristics are:

- it is produced from water which is readily available and cheap
- it is clean, odourless, tasteless and sterile
- it is easily distributed and controlled
- when condensed, it gives up heat at a constant temperature
- it has a high heat content
- it can be used to generate power and then to provide heating

Steam may be produced in any one of the three following conditions:

- wet steam
- dry saturated steam
- superheated steam

When steam is generated from water, it may be taken through several heating stages to produce steam in the condition required.

First, the water receives sensible heat to bring it to the temperature at which steam begins to form. This is 100°C (212°F) at atmospheric pressure or a higher temperature if the pressure is increased.

Secondly, the water at the temperature and pressure of steam formation receives latent heat to evaporate the water into steam at the same temperature and pressure. The heat required is approximately 2250 kJ/kg at atmospheric pressure. If the water is completely vapourised it is called "dry saturated steam". If, however, the steam contains a proportion of unvaporised water in suspension, it is "wet steam". Wet steam has only received a fraction of the latent heat required to turn all its water content into dry saturated steam.

Finally, if required, the dry saturated steam may be given sensible heat to raise its temperature still further and produce "superheated steam".

288

The condition of wet steam is specified by its "dryness fraction". This is the ratio of actual steam to wet steam.

$$\text{dryness fraction} = \frac{\text{weight of actual steam formed}}{\text{total weight of wet steam}}$$

For example, if 1 kg of wet steam has a dryness fraction of 0.8 then it contains:

0.8 kg of actual steam
0.2 kg of water
or   800 g of steam
200 g of water

This 1 kg of wet steam has, therefore, only received 0.8 (or 80%) of the total amount of latent heat required for its vaporisation. That is:

0.8 × 2250 = **1800 kJ**

138   Types of Boiler

Boilers are generally sized by their steam output or "evaporating capacity". This is the quantity of water in kg/h (lb/h) which can be evaporated from and at 100°C (212°F). Working pressures vary with the type of boiler and the process requirements. They range from 350 mbar (5 lbf/in²) for small gas boilers, up to about 160 bar (2300 lbf/in²) in modern power stations.

Boilers are classified as either "fire tube" or "water tube" types.

Fire tube boilers, or "shell" boilers as they are called, are generally cylindrical in shape. They have burners firing through tubes which pass through the water in the shell from one end to the other.

Fire tube boilers may be subdivided into either "single-pass" or "multi-pass" types.

Single Pass Fire Tube Boilers

These boilers have one set of fire tubes with burners at one end and the flue at the other, Fig. 1. They may be mounted with the shell vertical or horizontal. The burners fire directly into each tube and are normally fanned draught on the horizontal boilers and natural draught on the vertical boilers. The boilers are designed for gas firing and have outputs of from 36 kg/h (80 lb/h) up to about 360 kg/h (790 lb/h). A typical vertical boiler is described in section 140. Boilers of this type are commonly used by dry cleaners and clothing manufacturers to operate steam irons and Hoffman presses.

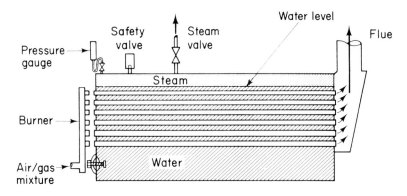

Fig. 1   *Single pass fire tube boiler*

Multi-pass Fire Tube Boilers

This boiler usually has a single main combustion tube with other sets of tubes passing the hot gases to the front of the shell and back again, Fig. 2. It is generally available as a packaged unit complete with burners and controls. It may be oil, gas or dual-fuel fired. Outputs may be up to 23 000 kg/h (50 500 lb/h) at pressures up to 18 bar (260 lbf/in^2). Boilers over 46 000 kg/h (101 500 lb/h) output generally have two combustion tubes.

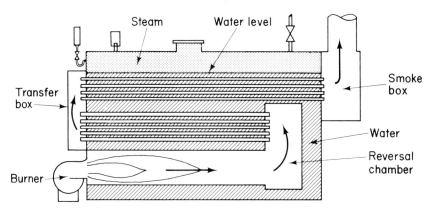

Fig. 2   *Multi-pass fire tube boiler*

Water Tube Boilers

These are the large, high pressure boilers used for industrial or power generation purposes. The hot gases from the burners pass around vertical banks of tubes containing the water. The boilers are roughly rectangular in shape and the tubes are connected to a water drum at the bottom and to a steam drum or manifold at the top. There is usually a superheater above the main combustion chamber. Outputs are generally above 20 000 kg/h (44 000 lb/h). Because of economic factors these boilers have generally been fired by pulverised coal or oil. Some have been converted to gas firing on an interruptable basis.

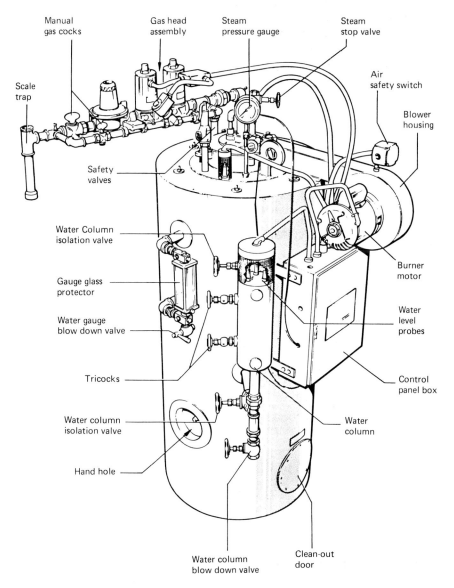

Fig. 3   Small steam boiler

Coil Type Boilers

These are a form of water tube boiler with the water contained in sets of coils. The burner fires down into the centre of the inner coil and the products pass around the outer layers of coils. They are low water capacity boilers and produce small quantities of steam quickly. The outputs may range from 200 kg/h (440 lb/h) to about 3 000 kg/h (6600 lb/h). They are either gas or oil fired by packaged burners.

Other Boilers

In addition to those described, there are small, gas fired boilers which are used to provide wet steam for bakers' ovens, Fig. 3. This "flash" steam is introduced into the oven for about 15 minutes during baking to give the loaves or rolls a rich brown, crusty surface.

The boilers are vertical and usually wall-mounted with outputs up to about 90 kg/h (200 lb/h). They are used principally by small, specialist bakers. Working pressures are in the region of 2 bar (30 lbf/in^2).

§139   Safety Regulations

Factory Acts

All steam boilers and their installations must conform to the requirements of The Factory Act (1937) and subsequent legislation.

This specifies that boilers must be fitted with various "mountings" which include:

- a suitable safety valve
- a suitable stop valve connecting the boiler to the steam pipe
- a steam pressure gauge, showing the maximum permissible working pressure
- at least one water gauge of transparent material, fitted with a guard if the working pressure exceeds 2.75 bar (40 lbf/in^2)
- where there are two or more boilers, each should bear a clearly visible plate showing a distinguishing number.
- unless a boiler is externally fired it should have a suitable fusible plug or an efficient low water alarm device.

All boilers, together with their mountings, must be inspected, both internally and externally, at least once every fourteen months, by a competent person. This person is usually the inspector employed by the insurance company.

Insurance

Steam boilers should be insured. Requirements vary, but premiums are usually related to boiler output. In addition to the annual inspection, most companies check the external mountings at six-monthly intervals.

The Insurance Companies' Associated Offices Technical Committee (AOTC) specifies the equipment which should be fitted to automatically controlled steam boilers as a condition of insurance. For boilers operated without supervision this includes:

- water feed control
- low water cut-off and audible alarm
- flame protection equipment.

Additionally, it is recommended that shell boilers and others having a perceptible water level which operate without regular supervision should be provided with an independent and separately operated device to cut off the fuel and air supply to the burners and to operate an audible alarm when the water level reaches a predetermined low position. The device should require manual resetting to reinstate the flame.

SMALL GAS FIRED STEAM BOILERS

140   Construction

This chapter is principally concerned with small, vertical, single pass fire tube boilers designed for gas firing. A cross section of a typical boiler is shown in Fig. 4.

The boiler consists of a welded steel shell containing tubes of up to 50 mm (2 in) diameter fitted between the end plates. The shell contains water up to $\frac{2}{3}$ of its height and steam in the top $\frac{1}{3}$. The water level is indicated by a gauge positioned between steam and water usually protected on three sides by glass plates. The shell has "mud doors" which are removed periodically to clean out any deposits and inspection doors to allow the fire tubes to be examined.

The burner is fitted with jets which are positioned to fire up each of the tubes. Baffles or "retarders" are suspended in the tubes to assist heat transmission to the water. On some boilers the burner assembly is mounted on a swivel. When the burner is swung out, the gas rate is automatically reduced to the minimum necessary for

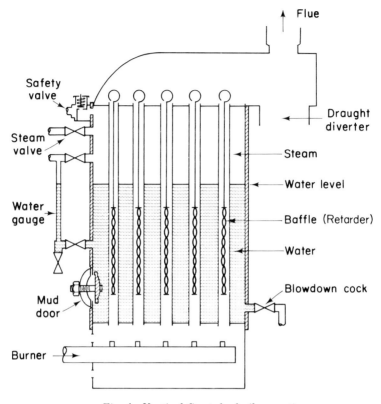

Fig. 4 Vertical fire tube boiler, section

manual ignition. On returning the burner the flames receive their full gas rate.

The top cover collects the products of combustion and directs them to the flue. A draught diverter is attached or incorporated.

§141 Controls

The gas controls fitted to small gas fired boilers include:

- main gas control cock
- gas pressure governor
- pressurestat
- low water gas cut off and alarm
- flame protection device
- low pressure gas cut-off
- interlocking pilot and main gas taps or burner swivel and cock for manual ignition

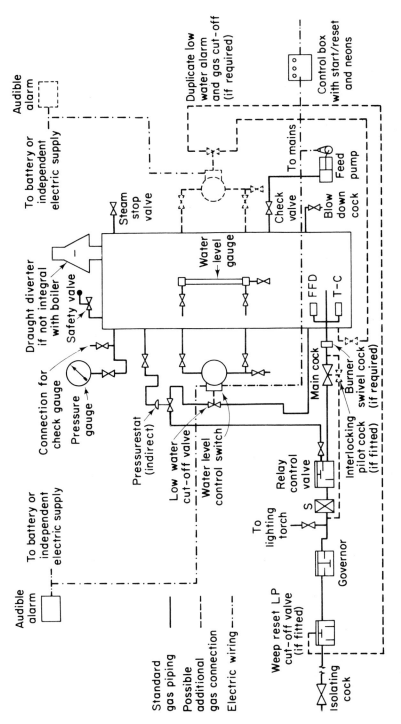

Fig. 5 Control layout

The steam and water feed controls include:

- main steam valve
- safety valve
- pressure gauge
- water gauge
- injector
- automatic feed control
- feed pump
- blow down cock

A diagrammatic layout of the controls on a typical boiler is shown in Fig. 5. Details of the various controls are as follows:

Main Gas Control Cock

A cock should be fitted upstream of all the gas control devices to enable them to be isolated for servicing or repair.

Gas Pressure Governor

The gas supply should be governed to the required inlet pressure by a constant pressure governor.

Pressurestat

A direct-acting pressurestat was described in Chapter 6 and the small boilers used in bakeries normally have a similar device. The larger, fire tube boilers generally use an indirect or an electrical pressurestat. The former may be fitted on the weep pipe of an ordinary relay valve, Fig. 6, or be integral with the relay valve, Fig. 7. In the type illu-

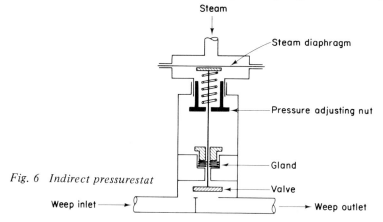

Fig. 6   Indirect pressurestat

strated, steam pressure from the boiler enters the device at A and is
fed to the bellows B. As the pressure rises it compresses the bellows
against the tension of the springs C. The gas valve meanwhile is held
open by gas pressure under the diaphragm F.

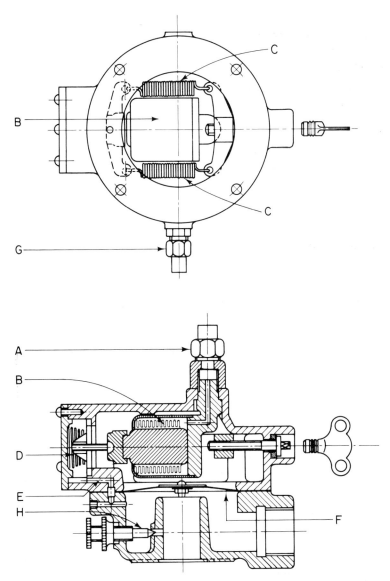

Fig. 7 Indirect pressurestat with integral relay valve

When the required steam pressure is reached, the bellows pushes open the weep valve D. This allows gas to pass from below the diaphragm to the upper chamber through weep E. Because gas enters the chamber faster than it can escape through a small orifice at G, the pressure above and below the diaphragm are equalised and the diaphragm falls, closing the gas outlet. A small amount of gas is allowed to pass to the burner through the needle valve H, to maintain the flames.

When steam pressure falls, the bellows expands, the weep valve closes and the pressure above the diaphragm is dissipated to atmosphere through the weep at G. The inlet pressure below the diaphragm lifts the valve and restores the main gas flow.

The pressure at which the pressurestat will shut off the gas is set by adjusting the tension of the springs C.

Low Water Cut Off and Alarm

The water level in a boiler is usually controlled automatically by means of a ball float. This can be made to operate mercury switches and a gas valve in order to:

- control an electrically driven water pump to feed water into the boiler
- activate an audible alarm
- open or close a small gas valve in the weep line to the pressurestat

Water may also be fed to the boiler by a hand pump or a steam injector feed. One type of ball float control is shown in Fig. 8. The float is pivoted in a fulcrum plate at A and the float arm is sealed off by a flexible bellows B which acts as a gland. The outer end of the arm rises when the float falls and tilts the upper mercury switch C, to activate the pump motor. Further movement will operate the gas valve at E and shut off the gas through the pressurestat by equalising the pressures above and below the diaphragm. The lower mercury switch D is tilted and should make contact just before the float reaches its lowest level. This activates the alarm. The alarm may be either mains or battery operated. The pump switch should break contact just before the float reaches its highest level.

A duplicate low water alarm and gas cut-off is usually required as a condition of insurance. This could be as previously described, but without the upper mercury switch. Alternatively a device may be used with one or two electrodes which complete a circuit when they are immersed in water. A single electrode control is shown in Fig. 9. If the water level falls below the bottom end of the electrode the sensing circuit is broken, the solenoid valve closes and the alarm is activated.

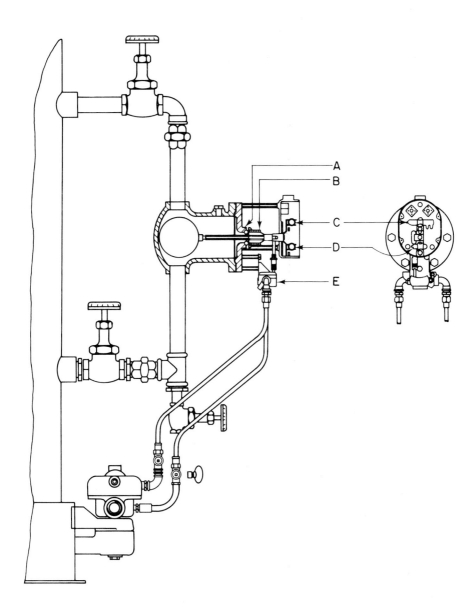

Fig. 8   Ball float water feed control

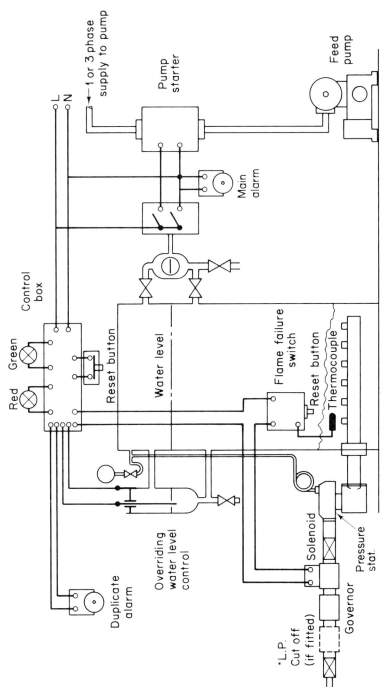

Fig. 9   Electrical controls, single electrode

*The L.P. cut off should be fitted downstream of the safety shut off valve unless it has an internal weep reset.

Flame protection device

The application of a thermo-electric flame failure switch is also shown in Fig. 9. It consists of a switch which can be closed manually by pressing and holding in a reset button. When the thermocouple is heated it energises a magnet which holds the switch in the closed position. If the flame fails, the magnet ceases to be energised and the switch opens. The device is wired in series with the gas solenoid valve.

Modern systems use flame rectification with a semi automatic control box (Chapter 5).

Low Pressure Cut-Off Valve

On boilers which are not electrically controlled it has been the practice to fit a weep reset low pressure cut-off valve immediately downstream of the main gas cock. The cut-off may be fitted with a weep line so that it can be operated by a low water level control.

Interlocking Taps or Burner Swivel

On boilers with manual ignition it is necessary to ensure that the main gas is not turned full on for lighting purposes. This can be done

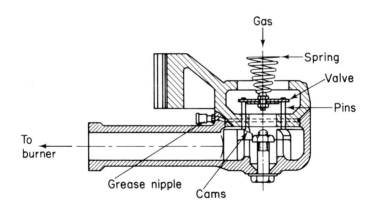

Fig. 10   Burner swivel with reducing valve

by having interlocking pilot and main gas taps so that the pilot must be turned on before the main gas tap can be turned. Alternatively a burner swivel with a reducing valve may be used, Fig. 10.

Main steam valve

This is a screw down stop valve fitted directly on to the boiler, near to the top. The main steam supply is connected directly into the valve.

Safety Valve

A totally enclosed, spring loaded safety valve, Fig. 11, is fitted towards the rear of the boiler near the top. The valve is protected from interference by a padlock.

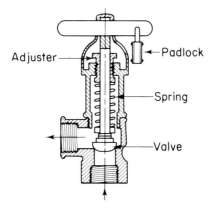

Adjuster ─────── ← Padlock

─── Spring

─── Valve

Fig. 11   Spring-loaded safety valve

Pressure Gauge

A bourden tube gauge complete with syphon and cock is mounted facing the front of the boiler.

Water Gauge

This consists of a stout glass tube held at top and bottom by packing glands, Fig. 12. It is fitted with steam and water inlet cocks and a drain cock. The tube is protected on three sides by heavy plate glass panels.

Injector

This is a device for feeding water into the boiler by the suction created when steam passes through a small jet, Fig. 13. It is brought into operation by first opening the feed check valve shut off cock and the suction cock and then opening the injector steam valve fully and quickly. Once the injector has been brought into use it can normally be controlled by the steam valve only.

Although it may only be used as a stand-by for an electric feed pump it should be operated daily to maintain it in working order.

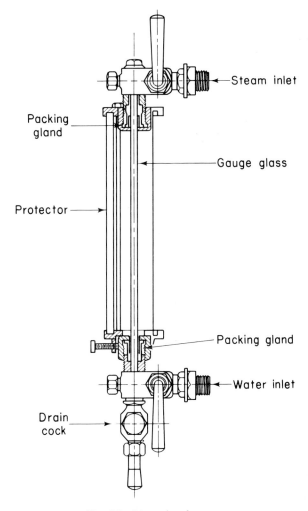

Fig. 12   Water level gauge

Automatic Feed Control

This may be part of the low level control device as already described.

Feed Pump

This may be either a simple hand-operated pump or an electrically driven feed pump as shown in the installation diagram in section 142. Whether by automatic or manual means, water must be fed into the boiler frequently and regularly to maintain the level at the middle of the gauge glass.

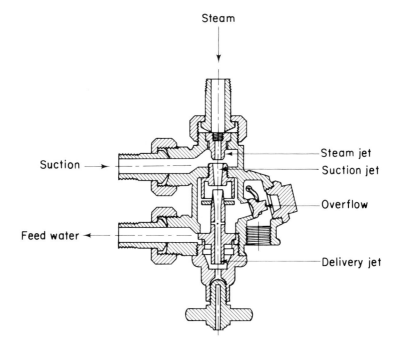

Fig. 13   Injector

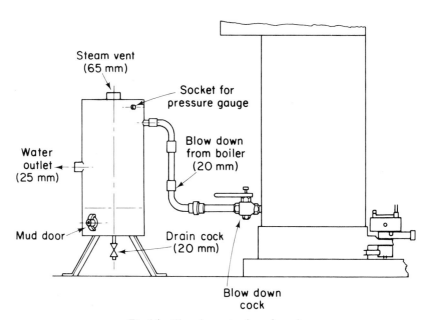

Fig.14   Blowdown tank and cock

Blow Down Cock

A special blow down cock is mounted on the boiler and connected to
the supply to a blow down tank, Fig. 14. The cock is usually a full-
way, lubricated plug type. It is fitted at low level at the rear of the
boiler and its function is to prevent any accumulation of mud in the
bottom of the boiler. This is done by regularly opening the cock
when the boiler is under steam. The cock should normally be kept
open until the water level in the gauge glass has fallen by at least
25 mm (1 in).

§142   Installation

The manufacturer's installation instructions should be consulted
before beginning any work.

The boiler should be mounted firmly and level on a concrete base
about 100 mm (4 in) thick. It should be sited to provide access to all

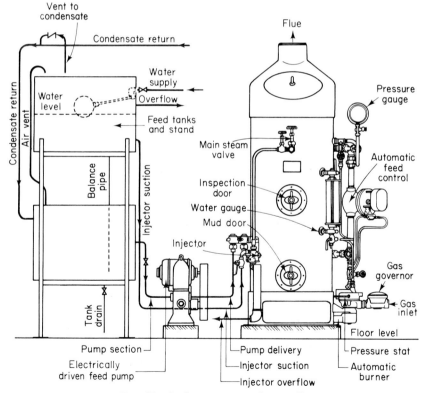

Fig. 15   Boiler water supply installation

ancillary equipment for servicing or repair. If the boiler has a swivel burner, space must be allowed for it to be swung out to its full extent.

Typical connection sizes are as follows:

- gas supply   25 to 50   mm   (1 to 2 in)
- flue pipe   75 to 200 mm   (3 to 8 in)
- steam supply 15 to 32   mm   ($\frac{1}{2}$ to $1\frac{1}{4}$ in)
- water supply
  - ball valve   15 mm ($\frac{1}{2}$ in)
  - injector feed   15 mm ($\frac{1}{2}$ in)
  - pump feed   20 mm ($\frac{3}{4}$ in)

The flue should be fitted with a draught diverter if this is not already integral with the boiler. It should be run in accordance with the recommendations given in Volume 2, Chapter 5.

The floor must be of adequate strength to support the weight of the boiler which, when filled with water may be from 250 to 1000 kg (0.25 to 1 ton).

The cold water supply may be taken from a boiler feed tank fed by a cistern as in Fig. 15.

§143   **Commissioning**

After testing gas and water supplies for soundness and purging, fill the boiler with water to the normal working level. This may most easily be done by removing the safety valve. Replace the safety valve as soon as the boiler is full and before lighting the gas.

Light the boiler in accordance with the manufacturer's instructions. Check:

- burner pressure
- burner bypass rate
- flame picture, flames correctly positioned below firetubes
- gas rate at the meter, if necessary
- flue for spillage at the draught diverter
- operation of flame failure device
- low pressure cut off valve

Vent the air from the top of the boiler shell through the water gauge. Close the gauge water cock and open the drain cock. Leave the drain open until steam is emitted. Close the drain and open the water cock. In the operating position all cock handles are normally vertical.

Turn on the main steam valve slowly and allow steam to pass to the plant. Check that the pressurestat shuts down the gas at the required working pressure. Check the automatic water feed and low level cut off and alarm. Switch off the power to the electric pump and allow the water level to fall so that the alarm sounds and the gas cut-off operates. Switch on the pump. The water level should return to normal, the gas supply be restored and the alarm switched off. The pump should stop at the normal working level. Check any duplicate cut off and alarm system.

Finally check the blow down cock and the safety valve and operate the injector system.

To shut down the boiler turn off the main gas cock and the steam valve. To prevent the alarm operating when the boiler cools down and the water level falls, it is necessary to fill the boiler to a high level when shutting down. This can be done either by using the injector or the hand pump. After filling, close the injector steam and suction valves. If an electric pump is fitted shut off the feed check valve cock and the pump inlet cock and switch off the power supply.

144 Servicing

Routine Checks

In addition to the normal periodic servicing, a number of regular daily checks and routine operations should be carried out as follows:

In use, ensure that the boiler operates at the correct working pressure by observing the pressure gauge.

Check that the water level is maintained at about the middle of the gauge glass.

Water Gauge

Check regularly that the connections are clear.

Shut off both the steam and water cocks and open the drain cock. Then open the other two cocks in turn to let steam and water blow through their connections and out of the drain. Finally close the drain cock and re-open the steam and water cocks.

Automatic Water Feed Control

This should be flushed out daily to ensure that the connections and the float chamber are clear of deposits. The control is connected to the boiler in a similar manner to the water gauge and a somewhat similar procedure should be followed.

Close the steam and water valves and open the drain valve slowly. Open the water valve to flush out the connecting pipe and then close the drain valve. Close the water valve, open the steam valve and then open the drain valve slowly to clear the steam connecting pipe and the chamber.

Finally, close the drain valve and open the water valve to restore the valves to their normal operating position.

Injector

This should be operated at least once a day to maintain it in good working order. It operates by opening, in turn, the following cocks or valves:

- feed check valve cock
- suction cock
- injector steam valve

If water issues from the overflow, the suction cock should be gradually closed until this stops.

Blow Down

The blow down cock should be opened regularly and frequently to prevent deposits building up around the bottom of the fire tubes. If the cock has a lubricated plug it should be re-packed regularly with lubricant and the plug operated to distribute the lubrication. This should be done when the cock is hot.

Safety Valve

The safety valve should be tested frequently to ensure that the valve is free to move and not choked with scale.

§145 Major Servicing

A full service and clean out is usually timed to coincide with the visit from the boiler inspector.

The operations to be carried out should include:

- question the operator and examine the boiler for visible faults
- shut down the boiler and isolate any electrical supplies
- remove and clean the burners, replace jets as required
- remove the top cover and tube baffles, brush out the fire tubes, clean and replace the baffles and the cover
- drain the boiler, remove the mud-doors and clean out the interior
- replace the doors, refill the boiler and check for leakage
- check gas taps and ease if necessary
- check gas soundness

- light and reposition the burner
- check burner pressure and flame picture
- check burner bypass rate
- check flue for spillage at the draught diverter
- check all controls and adjust as necessary
  - flame failure device
  - low pressure cut off
  - pressurestat
  - automatic feed and low water cut off and alarm
  - any duplicate alarm and cut off system
  - water gauge
  - injector
  - blow down
  - safety valve
- check all steam and water cocks operate satisfactorily

146   Fault Diagnosis and Remedy

Faults on governors, low pressure cut off valves and flame failure devices were dealt with in Volume 1.

Two common faults, failure of the electrical or gas supplies, would have the following consequences.

Electric Supply Failure

This would close any solenoid valve and the feed pump would cease to operate. The boiler would shut down. In the case of a boiler with a feed pump but no solenoid valve, the water level would fall until the gas cut-off operated to shut down the boiler. Any independently supplied low water alarm would be activated.

Gas Supply Failure

In this event the flame failure device and the low pressure cut off would close and remain closed until manually reset. The boiler would shut down.

Other Common Faults

Any major leakage would obviously require the boiler to be shut down immediately. However, small leakages from glands, joints or any other source should also receive immediate attention to prevent dangerous situations developing.

Faults on specific devices are as follows:

Safety Valve

Safety valves often develop slight leakages and for this reason it is tempting to leave them alone when all seems well. Testing the valve to ensure that it is free may start it leaking but the test is essential to ensure that the valve will operate in an emergency. Any slight leakage should be stopped immediately. Shut down the boiler, dismantle and clean the valve, regrinding the valve and seating. Reassemble, test and adjust to the correct pressure.

Injector

The injector failing to operate may be due to:

- feed check valve not seating properly
- leaks in suction or delivery pipes
- mud or scale in the injector
- injector overheated due to leaking steam valve or several unsuccessful attempts to operate
- feed water too hot

The injector and the feed check valve may be isolated for cleaning by closing the injector steam valve and the feed check valve shut off cock while the boiler is still under pressure.

Automatic Feed Control

The operation of the float can be checked by manually operating the switch levers between the float chamber and the switch box. The switches are adjustable by altering their angle on the switch carrier. The gas cut off valve is adjusted by rotating the knurled sleeve on the push rod after first slackening the lock nut. The float may be cleaned or renewed while the boiler is under pressure by closing the steam and water valves and opening the drain valve. Then isolate the electrical supply and close the pressurestat weep cocks. The switch gear and the float chamber can then be dismantled. In this situation the water level in the boiler must be maintained by means of the injector or the hand pump.

If the bellows gland develops a leak, shut the steam and water valves to stop steam or water escaping and open the drain valve. The water pump can be controlled by the main switch and the pressurestat kept operative by closing the relay weep cocks. There is then no protection against low water level and great care must be taken to ensure that the boiler operates safely. The gland must be replaced before the boiler can be left unsupervised.

If, for any reason, the water level should become too low and not be visible in the glass, turn off the gas supply immediately. Allow the boiler and its ancillary equipment to cool down naturally and completely before admitting any cold water. If cold water is introduced into an empty, pressurised boiler it could cause an implosion or leaks around the fire tubes due to rapid contraction. In either case the boiler would be damaged beyond repair.

Pressurestat

On any type of pressurestat a leaking steam bellows will result in excessive steam pressure, since the main gas valve will not shut down. This may result in the safety valve blowing off if the higher pressure is not noticed quickly.

With indirect pressurestats, faults on the weep line, the indirect controls or the diaphragm may have different effects on the various types of relay valve or main control valve.

On an ordinary relay valve controlled by an indirect pressurestat, the gas valve will stay open when:

- the pressurestat bellows is leaking
- the pressurestat valve does not close or close completely
- the weep line is broken or leaking
- the relay valve diaphragm is leaking
- the relay valve weep jet is choked

The gas valve will stay closed when:

- the pressurestat valve does not open
- the weep line is blocked

On the pressurestat shown in Fig. 7, the gas valve will stay open when:

- the pressurestat bellows is leaking
- the weep valve is closed or sticking

The gas valve will stay closed when:

- the weep valve does not close or close completely
- the main diaphragm is leaking

In all cases when dealing with faults on any device it is advisable to consult the manufacturer's instructions before carrying out any work.

CHAPTER 9

Overhead Heating

Chapter 9 is based on an original draft by Mr. B. Gosling

---

§147 Introduction

Industrial and commercial buildings may be heated by either radiant or convected systems. In industrial premises in particular, floor space is usually at a premium and fairly high, open roofs offer a location for heaters, emitters and air ducts.

So industrial space heating is often by overhead heating.

In commercial premises suitable heaters or ducting can be accommodated in suspended ceilings.

The types of heater in use include:

- radiant heaters   —high temperature, luminous radiant panels
  —radiant tube heaters
  —air heated radiant tubes
  —steam or water heated radiant systems

- convection heaters—unit air heaters
  —direct fired air heaters
  —indirect fired air heaters
  —make-up air heaters.

Radiant heaters are normally mounted at high level on the walls or suspended from the roof trusses.

Of the convection heaters, unit air heaters are similarly mounted and make-up heaters are roof-mounted. The direct fired air heaters may be fitted overhead, in the form of horizontal or inverted heaters, or stand on the floor. All convection heaters may be fitted with or without ducting.

RADIANT HEATING

148 Radiation

Radiation was dealt with in Volume 1, Chapter 9.
It was established that:

$$Q = \sigma E A K^4$$

where Q = heat energy
  σ = Stephan-Boltzman constant (5.67×10^{-8} W/m^{2}K^4)
  E = emissivity (black body = 1)
  A = surface area
  K = absolute temperature, Kelvin.

Radiation also obeys the inverse square law, that is, the intensity of radiation is reduced in inverse proportion to the square of the distance between the emitter and the receiving surface, the total energy remaining constant Fig. 1.

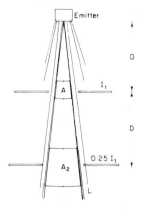

Fig. 1   Inverse square law

The following additional points have a bearing on the design of overhead heating schemes.

When rays from two or more sources are projected on to a surface, the total energy received is the sum of the output from each source, Fig. 2.

Infra-red rays may be projected in any direction without affecting the amount of energy received on the absorbing surface. Each of the surfaces at (a), (b) and (c) in Fig. 3 will receive an identical amount of heat.

The maximum intensity of radiation from a flat panel is received

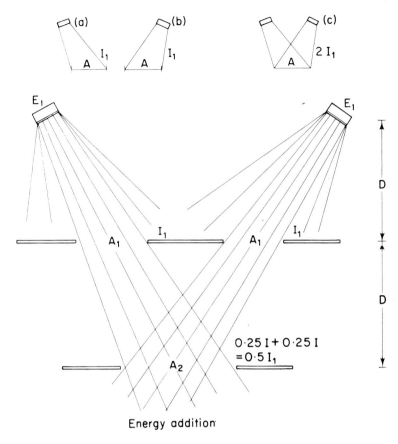

Fig. 2   Energy addition

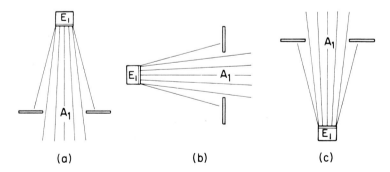

Fig. 3   Direction of emission (a) downwards, vertically
(b) horizontal (c) upwards, vertically

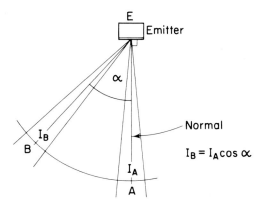

Fig. 4   Reduction in radiant intensity with reduction in angle of emission

at a point on a perpendicular line from the centre of the panel. That is, point A in Fig. 4. This line at right angles to the panel is called the "normal". When heat is received at points which are at angles of less than 90° to the panel's surface, the intensity of radiation is reduced. When the angle with the surface becomes zero, the amount of radiation is zero. The intensity of radiation at any point B is actually the intensity of $A \times$ cosine α where α is the angle between EB and

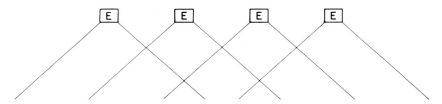

Fig. 5   Arrangement of heaters to give even heat distribution

the normal EA, Fig. 4. This means that a single heater will not heat an area evenly and several heaters may be necessary to give an acceptable heat distribution, Fig. 5.

Gas fired heaters radiate on nearly all the infra-red wavelengths, from 0.8 to 500 micrometres. However, for any particular heater, the emission is at a maximum for a particular wavelength. The maximum emission is related to the temperature of the emitter in accordance with Wien's Law, discovered by a German physicist. Actually, the wavelength for maximum emission,

$$\lambda_{max} = 2884 \times 1/K \text{ micrometres}$$

where K is the absolute temperature of the emitter.

Luminous heaters emit rays of a relatively short wavelength, λ_{max} = 2.5 micrometres. They operate at temperatures of from 800°C to above 1000°C (1470°F to above 1830°F).

Non-luminous heaters emit rays of longer wavelengths, λ_{max} = 4 micrometres. Their surface temperatures are usually between 300 and 600°C (570 and 1110°F). These conditions produce weaker rays and more gentle heating.

§149   Application of Radiant Heating

Radiant heating has advantages when used in the following situations:

- enclosed areas with a height of more than 6 m (18 ft)
- a partially enclosed area, for example, a loading bay or sports stadium
- selected areas of a large enclosure, for example, a work bench or a particular machine in a large production area
- areas requiring heat for short periods

Figures 6 and 7 show a workshop. In Fig. 6 it is heated by two air heaters and in Fig. 7 it is heated by overhead radiant heaters.

As the air is warmed, in Fig. 6, it becomes less dense and rises above the colder air. So there is comparatively cooler air at ground

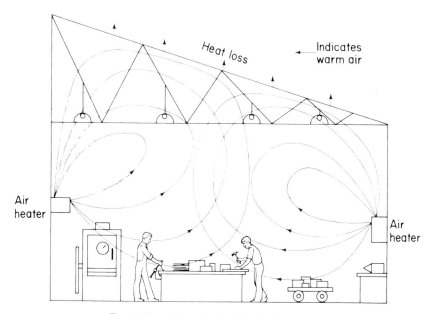

Fig. 6   Workshop heated by air heaters

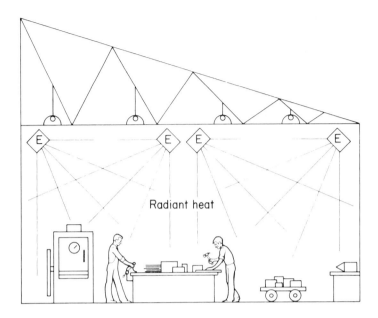

Fig. 7   Workshop heated by radiant panels

level and warmer air under the roof. The temperature gradient between floor and roof may be up to 11 deg C (20 deg F) and it varies in proportion to the height of the space and the mixing achieved by the heater. This can result in a high heat loss from the roof. Large air changes will also increase running costs.

The air is circulated through the heaters by fans. High air velocity can cause draughts near return air grilles and may stir up dust in some environments.

In Fig. 7, the radiant panels emit infra-red rays which are immediately absorbed by the people and equipment at ground level. Static air in contact with the roof is generally at a slightly lower temperature than that in the working area, which gains a little heat from the contents of the workshop. The air plays no real part in the transmission of energy and the air temperature may be quite cool without detracting from comfort. Temperatures may be 6 deg C (11 deg F) below those required with convection heating.

Luminous radiant heaters rapidly reach their operating temperature. So people in the path of the radiation are quickly made comfortable although the air and the building structure are still cool.

If a radiant heating system is used for 8 to 10 hours per day, it is generally only necessary to turn it on at the time that heating is required. After a weekend break an initial warm-up period is required, particularly in cold weather.

The directional nature of radiant heaters allows them to be used in spaces which are not totally enclosed. Figure 8 shows a typical loading bay. The partially enclosed area should be screened from high winds.

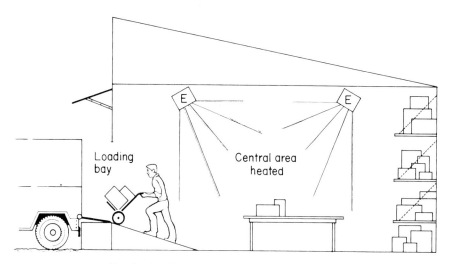

Fig. 8   Loading bay heated by radiant panels

§150   Luminous Radiant Heaters

These heaters utilise the radiant burners described in Volume 1 Chapter 3, usually with natural draught injectors. To give flame retention on natural gas a stainless steel gauze may be fitted about 5 mm ($\frac{1}{4}$ in) below the outer surface of the ceramic plaque. A typical heater is shown in Fig. 9.

One model uses a ring burner firing on to a tapered ceramic fibre gauze, surrounded by a stainless steel mesh, Fig. 10. A circular reflector of polished aluminium alloy concentrates the heat over a relatively small area.

Luminous heaters are not flued and the products of combustion are discharged directly into the premises. This is acceptable where the roof is high and there is a fairly high rate of air change. A certain amount of heat is carried away by the products and there is a possibility of condensation occurring on cold walls or steel work.

Heat outputs of about 15 kW (51 000 Btu/h) may be obtained from a typical heater, although heaters with greater outputs are available if required. A single plaque burner unit has a heat input of about 4 kW (13 500 Btu/h).

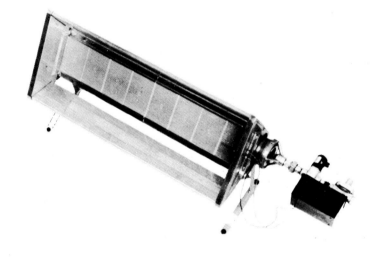

Fig. 9   Luminous radiant panel

Location

The number of panels required is determined by calculating the heat losses from the building as detailed in the C.I.B.S. Guide and in Volume 1 Chapter 9. Additional heaters may be required to cover loading bays and frequently opened doors.

Fig. 10   Circular luminous radiant heater

The information required when planning a heating scheme includes:

- use of the premises
- environmental temperature required
- type and area of walls
- window sizes and positions
- type and size of doors, including loading bays
- roof area and insulation
- floor construction and area
- height to ceiling, if fitted
- air changes per hour
- location of occupants, passageways, light fittings and equipment
- position of roof trusses and girders suitable for supporting heaters
- presence of air-borne contaminants, dust, paint spray, any solvents, or corrosive substances
- type of heater suitable and minimum fixing height specified by manufacturer

When planning the installation, check:

- position and size of gas meter and gas supply
- location voltage and phase of electricity supply, if required.

The closer that a radiant heater is to the ground the smaller the area heated but the greater the radiant intensity. An intensity of 79 W/m^2 (25 Btu/ft^{2}h) is normal for general heating at floor level. At working levels 158 W/m^2 (50 Btu/ft^{2}h) is comfortable but 236 W/m^2 (75 Btu/ft^{2}h) at head height should be regarded as a maximum value and is only required for exposed locations. Table 1 gives typical intensities and mounting heights.

151 **TABLE 1** **Radiant Intensities and Mounting Heights**

Radiant Intensity		Mounting Height	
W/m^2	Btu/ft^2	m	ft
230	73	3.7	12
183	58	4.3	14
104	33	4.9	16
91	29	5.5	18
85	27	6.1	20

Intensity for a given heater is at its maximum when the radiant surface is horizontal. Heaters may be fitted at up to 60° to the horizontal, Fig. 11. When the heater is angled the products of combustion escape more readily, allowing cooling air to pass over the heated surface.

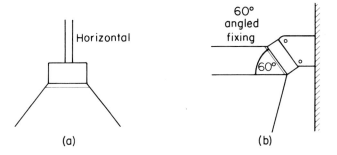

Fig. 11   Angle of mounting (a) horizontal (b) 60° to horizontal

When the number of heaters has been determined they should be arranged as evenly as possible over the area to be heated. Generally each point should be heated from at least two and preferably four directions to avoid shadowing, Fig. 12.

Examples of heater locations are given in Figs. 13 and 14.

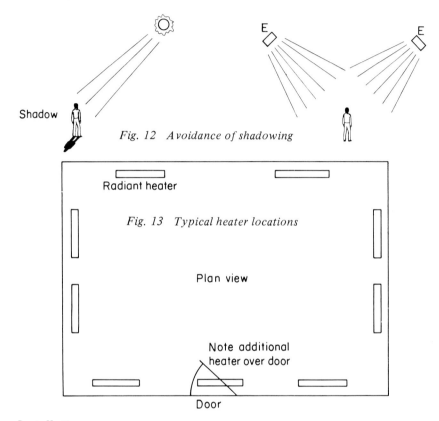

Fig. 12   *Avoidance of shadowing*

Fig. 13   *Typical heater locations*

Installation

A wide range of ignition and control systems may be used with luminous panels. These range from a simple lever cock with chains and a permanent pilot to semi-automatic and automatic systems which may incorporate thermostats and timers.

Heaters may be suspended from:

- the gas supply
- brackets
- chains.

Only lightweight heaters should be suspended from the gas supply. If the connecting pipe is more than 600 mm (2 ft) long a cup and ball

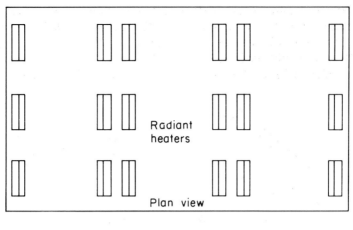

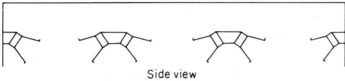

Side view

Fig. 14 Heater locations to cover large area

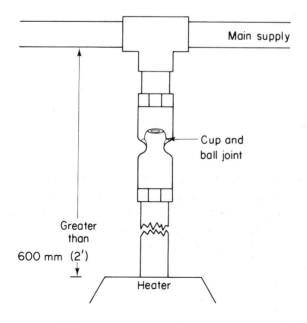

Fig. 15 Method of suspension

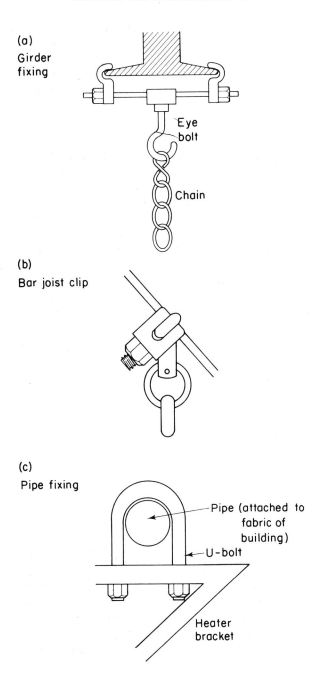

Fig. 16　Fixing devices (a) girder fixing (b) bar joist clip (c) 'U' bolt

joint should be fitted close to the main gas supply, Fig. 15.

Brackets are usually provided by the manufacturer. Some types are adjustable for either horizontal or angled mounting.

Chains may be supplied by the makers. They support the heater and prevent strain on the gas supply. The chains are secured to joists or girders by lugs, clips or screws, Fig. 16.

Servicing

Radiant heaters should be serviced annually prior to the heating season. Servicing should include cleaning injectors, burners and pilots and checking taps and ignition and control devices. In addition the reflectors should be cleaned and polished.

Dust from the atmosphere builds up on the rear face of the ceramic plaques and can cause local overheating and light-back. It may be removed by a fine jet of compressed air at a pressure of about 5.5 bar (80 lbf/in^2). This should be directed first into the holes in the plaque and then into the venturi and finally into the holes again. Care must be taken not to damage the plaque or dislodge the sealing material holding it to the burner body. Some manufacturers recommend washing the plaques using a detergent.

It may be necessary to replace damaged or badly linted plaques. This is best done on a bench. The new plaque must be sealed by the material appropriate to the particular model which may be metal strips, fireproof cement or insulating material.

152   Non-Luminous Radiant Heaters

Radiant Tube Heaters

These are heavier, longer and more robust than the luminous heaters. The principle of operation is shown in Fig. 17(a) and a burner at 17(b).

The burner fires into one end of a medium or heavy duty mild steel tube typically 65 mm ($2\frac{1}{2}$ in) diameter or a lighter gauge, 104 mm (4 in) diameter steel tube. At the other end is a fan which sucks the products of combustion through the tube and ejects them through a short flue. The fan suction also draws in the gas and combustion air to the burner. The tubes may be up to about 7 m (23 ft) long and their gauge and diameter vary with the manufacturer.

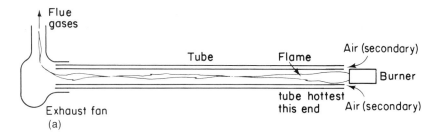

(a)

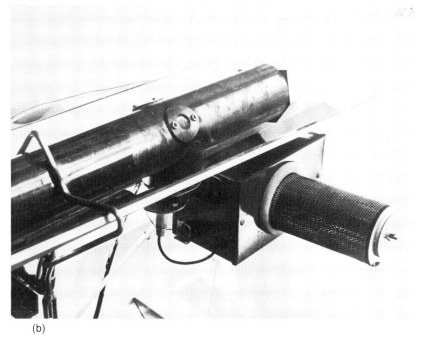

(b)

Fig. 17   Radiant tube heaters (a) principle of operation (b) heater

One manufacturer's heaters may be connected together in series with one large fan serving up to about 16 heaters, Fig. 18. Most models have one length of tube in a 'U' shape, Fig. 19.

The heaters are generally fitted with automatic controls. These normally incorporate a purge period, automatic spark ignition, flame monitoring and an air-flow switch. The gas supply is controlled by a governor or a zero governor so that a constant gas/air ratio is maintained irrespective of fluctuations in fan suction. Thermostats and timers may be connected to the system.

The radiant tubes are fitted with reflectors and the heaters are suspended from the roof trusses or girders, usually at a height of 3.5 to 4.5 m (12 to 15 ft).

Heat outputs of about 18 kW (61 500 Btu/h) are obtained from a 7 m (23 ft) tube. The final gas connection to the heater should be by armoured flexible tubing.

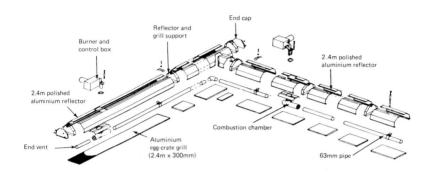

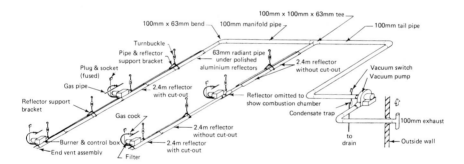

Fig. 18 Multiple tube installation

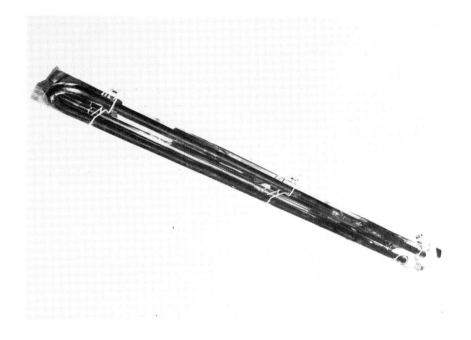

Fig. 19　Radiant tube heater

Air Heated Radiant Tubes

This system is shown in Fig. 20. It consists of a direct-fired air heater connected to a bank of circular metal ducts forming a closed circuit. The air heater and fan may be fitted in the roof space or mounted on the floor. A flue is taken to outside air.

The metal ducts are each about 600 mm (2 ft) diameter and may be in banks of up to 3 tubes. The tubes are fitted with an insulated top and with side shields to protect them from draughts and to

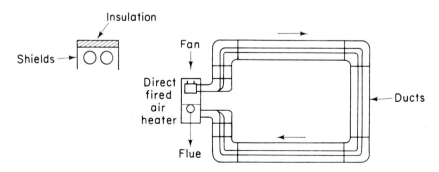

Fig. 20　Air heated radiant tubes

concentrate the radiation downwards. The ducts and shields are suspended from the roof trusses or girders.

Steam and Hot Water Radiant Systems

Medium or high temperature hot water at 120 to 180°C (248 to 356°F) or steam is piped from a central boiler through overhead panels or strips. These are usually mounted horizontally near to the roof.

The strips, which have superseded the unit panels, consist of continuous lengths of panel made of a plate heated by contact with one or more pipes. Heat output from a single-pipe strip may be about 525 W/m (550 Btu/ft h).

CONVECTION HEATING

153   Unit Air Heaters

Although this term has been used to cover all types of flued, forced convection air heater, it was first given only to those heaters which were fitted overhead. It is in this context that it is used here.

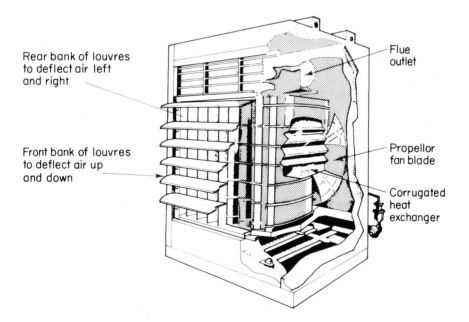

Rear bank of louvres to deflect air left and right

Front bank of louvres to deflect air up and down

Flue outlet

Propellor fan blade

Corrugated heat exchanger

Fig. 21   Unit air heater

Some unit air heaters were indirectly heated by steam or hot water piped from a central boiler. Air is blown by a fan through a finned water to air heat exchanger and directed on to the area to be heated by a bank of louvres at the front of the heater. This type is now largely being superseded by directly heated gas fired models operating in a similar manner. The products of combustion from the burners pass up through a tubular steel or clamshell heat exchanger to the flue. Air is blown by a centrifugal, or a propeller-type fan around the heat exchanger and out through the louvres at the front of the heater, Fig. 21.

Unit air heaters are suitable for installation in stores, halls and workshops where floor space is at a premium and quick heating up is required to eliminate cold spots. They are available for free air discharge or for use with ducting systems in the following variations.

1. Duct Model

A basic heat exchanger without fan or motor but with inlet and outlet spigots for installation in duct work with a separate forced convection system, Fig. 22(a).

2. Propeller Fan Model

Self contained unit with propeller-type or axial fan and adjustable louvres to direct the flow of warm air. A filter cannot be used, Fig. 22(b).

3. Centrifugal Fan Model

Similar to propeller type but with a centrifugal fan or blower, giving an increased air throw. Fitted with louvres for free discharge or a ducting spigot, Fig. 22(c).

4. Fan Compartment Model

With an enclosed centrifugal fan enabling a filter and return air ducting to be fitted, Fig. 22(d).

The heaters are fitted with controls similar to those used on domestic warm air heaters. These may include:

- ignition, often by permanent pilot
- flame protection, thermoelectric, usually in conjunction with a multifunctional control
- electric limit thermostat and fan control
- electric roomstat
- timer

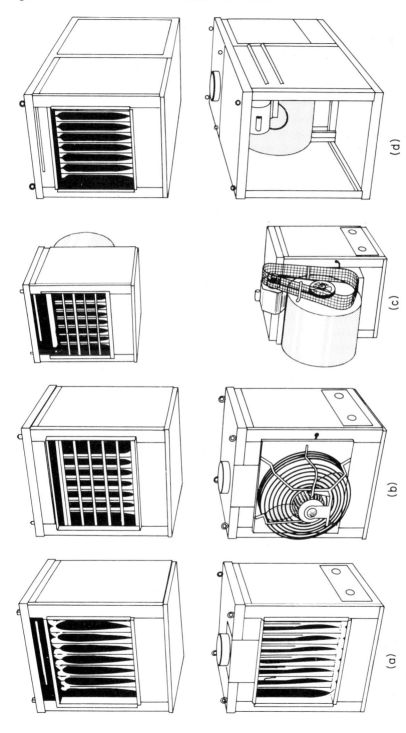

Fig. 22 Unit air heaters (a) duct model (b) propeller fan type (c) centrifugal fan type (d) fan compartment model

Some models have a summer/winter switch to allow the fan to be used to circulate cool air. The heat output of the average sized unit air heater is about 30 kW (102 500 Btu/h).

Installation

Heaters are usually suspended from roof trusses or girders although some models may be fitted on walls. Mounting height is about 2.5 m (8 ft) to the bottom of the heater with at least 1 m (40 in) clearance between the heater and a ceiling. Manufacturers' instructions should be followed in all cases. Care must be taken to ensure that the suspension is reliable and unaffected by vibration.

The louvres should be set so that warm air is directed on the area or personnel to be heated. Where several heaters are mounted in a room they should be positioned so that they all contribute to a general circulation of warm air around the enclosure. Fig. 23 shows typical locations for heaters to provide even heat distribution.

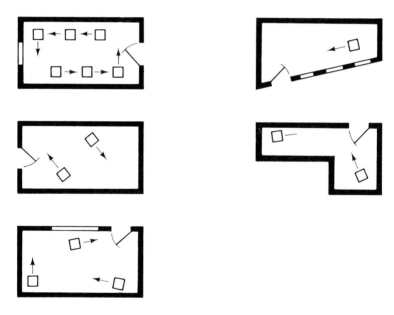

Fig. 23　Typical locations for unit heaters

Flueing

The heaters may be vented by open flues using stainless steel or asbestos cement flue pipe. Alternatively a fan diluted flue unit may be used when ordinary methods of venting are undesirable or impractical.

Examples are:

- when flue products discharge adjacent to an openable window
- it is not possible to pass flue through the ceiling or to fit a vertical flue from the heater location
- there is insufficient height between heater and ceiling

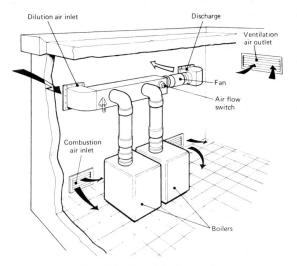

Fig. 24　Fan dilution unit

The fan dilution unit is shown in Fig. 24. It consists of a small centrifugal fan, air flow switch and low voltage relay. Although the layout is different the principle of operation is the same as described in Volume 2 Chapter 5. The unit serves two functions:

- it reduces the CO_2 concentration from about 5% to 1%
- it reduces the temperature of the flue gases from about 205°C (400°F) to 38°C (100°F)

This reduces "steaming" from the flue outlet.

Fresh air enters the unit through two parallel ducts on either side of the flue outlet. The products are conveyed to the outlet by a 125 mm (5 in) diameter metal fluepipe which may be connected in several ways.

The unit incorporates a thermal switch which ensures that the fan will continue to run while there are hot products in the flue.

After installation the unit should be tested by checking for spillage at the heater draught diverter. The damper in the fan unit should be adjusted so that no spillage occurs and air is drawn gently into the draught diverter. When set, the damper should be secured with a self tapping screw.

§154 Horizontal and Inverted Heaters

These large output heaters are modifications of floor standing models, as described in Chapter II, for installation overhead. They are normally supplied mounted in a steel framework, complete with a purpose-made servicing platform and guard rails, Fig. 25. Because of their weight and size care must be taken to ensure that they are securely supported, usually by the structural brickwork. Outputs range from about 44 kW to 1025 kW (150 000 to 3 500 000 Btu/h).

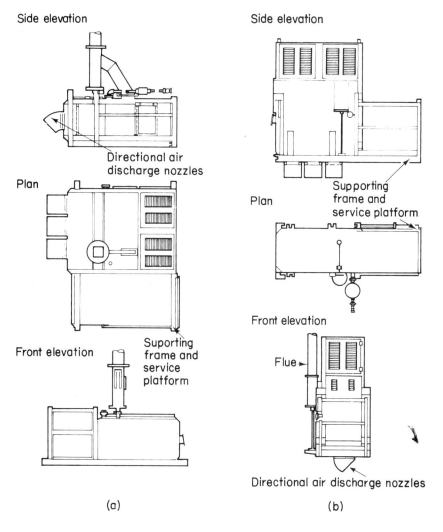

Side elevation

Directional air discharge nozzles

Plan

Front elevation

Suporting frame and service platform

Side elevation

Plan

Supporting frame and service platform

Front elevation

Flue

Directional air discharge nozzles

(a)

(b)

Fig. 25 Horizontal and inverted heaters (a) horizontal (b) inverted

Both types of heater may be fitted with several discharge nozzles which may be swivelled through 360° to supply heated air to all parts of the working area. Alternatively they may be connected to nozzle extensions or ductwork to convey warm air to cold spots or air curtains.

Most heaters are flued but some may be direct fired and allow the products of combustion to mix with the discharging warm air.

§155   Direct Fired Air Heaters

Because these heaters are not flued and discharge diluted products of combustion directly into the heated space it is necessary to obtain a waiver to the Building Regulations, from the local authority when an installation is proposed.

The heaters are often floor mounted but they may be suspended from the roof girders. Some models are available with weatherproof housings for mounting externally on a flat roof.

Fig. 26   Direct fired air heater

Figure 26 shows a smaller, cylindrical type. This is fully automatic with spark ignition and electrical controls. These include:

- air flow switch
- low gas rate start
- safety shut off valves
- limit thermostat
- summer/winter switch
- flame rectification or UV flame protection
- full sequence automatic control.

The axial fan is mounted at the rear of the body and may be operated for ventilation only.

Large rectangular models may also be mounted at high level and are used for normal heating or make-up air heating.

§156 Make-up Air Heaters

Air extraction systems are fitted in a number of commercial and industrial premises. For example, these are found in:

- restaurants and canteens—to ventilate the dining rooms and to remove cooking odours, steam and products of combustion from the kitchen
- garages—to remove flammable vapours
- factories—to remove unpleasant, toxic or flammable fumes.

As the polluted air is extracted, fresh air must be taken from outside the building to replace it and this replacement air is called "make-up air". Where rates of air extraction are high, special provision must be made for the entry of make-up air and for heating or cooling it according to its temperature. In this country, cooling is less common and it may only be necessary to blow in fresh air to

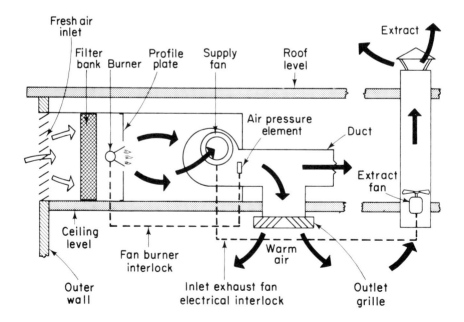

Fig. 27   Principle of make-up heater

obtain a cooling effect. However, cooling coils may be fitted to some type of heater.

The principle of the make-up air heater is shown in Fig. 27. Air is drawn in from outside the building by the suction caused by the circulating fan. The air enters through an inlet protected by angled louvres and a wire mesh grille to prevent the entry of birds, rain or leaves and is usually filtered before passing to the burner. Some of the air is used for combustion and the remaining air is heated. The products of combustion mix with the heated air and are propelled by the fan, at a slight pressure, to the outlet grilles.

One type of burner, Fig. 28, has perforated baffle plates attached to it which create a low pressure area in front of the burner ports

Fig. 28   Make-up air heater burner

into which combustion air is drawn. The profile plate, Fig. 27, controls the volume and velocity of the make-up air in conjunction with the fan.

The heaters are generally designed to provide a slightly greater volume of air than is being extracted. This allows a slight pressure to build up which distributes the warm air through the building and also prevents the ingress of cold draughts. The systems may either give general heating throughout the whole building, Fig. 29, or be ducted to a particular enclosure, like a paint spray booth or kitchen, to compensate for the air being extracted from that environment, Fig. 30.

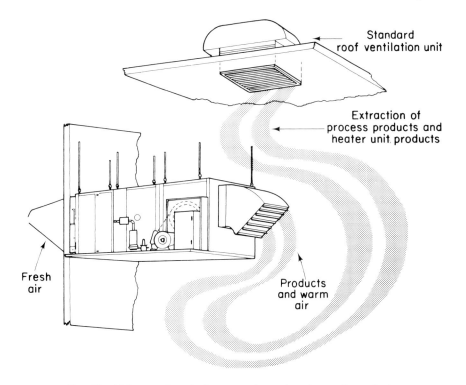

Standard
roof ventilation unit

Extraction of
process products and
heater unit products

Fresh
air

Products
and warm
air

Fig. 29   Make-up air unit for general area heating and ventilation

Heaters are available in a variety of sizes, typically from about 118 kW (402 500 Btu/h) to 880 kw (3 000 000 Btu/h). They are usually fitted in a roof space and have lifting lugs for easy handling and support, Fig. 31. Some models may be supplied with a weather-proof casing for external installation.

The controls supplied vary from one manufacturer to another. The heaters are generally fully automatic and a full sequence control is incorporated. This gives a purge period and has spark ignition usually to an interrupted pilot burner which shuts off when the main gas is established. Flame protection is by ultra violet detection or flame rectification.

The manufacturing, performance and safety requirements for air heaters with heat inputs of 60 kW to 2 MW (205 000 to 6 800 000 Btu/h) are under review by British Standards Institution. A new specification and a code of practice for installation will be issued in due course and it is likely that BS 3561 will be amended.

The fan on the heater is often interlinked with the extractor fan and will only run, on "winter" setting, when the extractor fan is already running. When the heater fan is running, an air flow switch

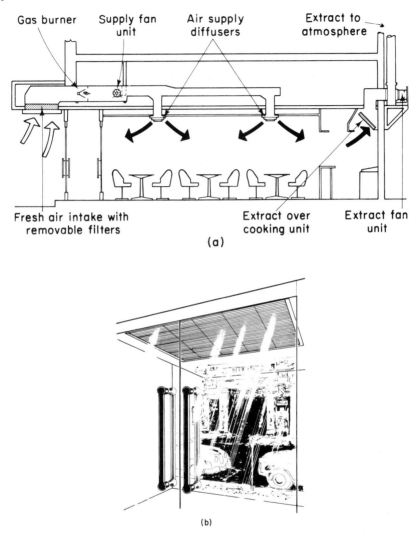

Fig. 30 *Compensating system replacing air extracted from canopy*

operates a second link with the burner controls. If either link fails, the sequence shuts down and must be manually reset.

Temperature control is usually by means of a thermostat in the exit duct operating on a modulating valve in the main gas supply. Turn down rates of 35:1 are possible.

Make-up air heaters with heat inputs above 60 kW (205 000 Btu/h) like similarly sized direct fired air heaters, contravene the Building Regulations with respect to Section M3. Tests have shown that the

Fig. 31   Make-up air heater

concentrations of CO_2 and CO in the heated air are considerably lower than the limits originally specified by the Health and Safety Executive. These are:

- CO_2 — 5000 p.p.m.
- CO   —   50 p.p.m.

The high dilution of the products of combustion by large volumes of make-up air results in a negligible concentration of CO and concentrations of CO_2 which are about half that specified as safe. Nevertheless, local waivers must be sought before installation can proceed.

CHAPTER 10

Air Conditioning

Chapter 10 is based on an original draft by Mr. A. J. Spackman

---

157 Introduction

The purpose of air conditioning is to provide an environment in which the temperature, moisture content and movement of the air are maintained at a level required to ensure the comfort conditions desired by the occupants. In some instances it is used to produce the special conditions necessary for a manufacturing process or a particularly sensitive piece of equipment, such as a computer.

In a fully controlled air conditioning system, air-borne dust from outside is eliminated by filtering. So dust is limited to that generated within the conditioned environment.

Where the environment must be closely controlled, it is essential that make-up air is only allowed to enter through the air conditioning system. So windows should not be opened and all external doors should be fitted with air locks. The environment must be maintained at a slight pressure so that any air leakage can only be outwards.

To maintain the desired conditions it is necessary to remove and replace some of the air in the controlled environment continually. The temperature and moisture content of the replacement air will be different from that of the air withdrawn and from that entering the unit from outside. The unit must continually process the make-up and return air to give the desired characteristics to the recirculated air.

158 Basic Principles

A number of the following principles were introduced in Volume 1, Chapter 4.

Moisture Content

Because part of the air conditioning process is concerned with controlling the moisture content of the air it is necessary to determine the amount of moisture contained.

This is usually measured in grams of water vapour (moisture) associated with 1 kg of dry air. It may also be in kg/kg or grains/lb. There is a limit to the quantity of moisture that air can contain and this limit varies with the temperature of the air. This limit is low when the air temperature is low and becomes greater as the temperature rises. When air contains the maximum quantity of moisture that it can hold at a particular temperature it is said to be "saturated".

Relative Humidity

"Saturation vapour pressure" is the pressure exerted by the maximum quantity of water vapour (moisture) that a given volume of air can contain under particular conditions of temperature and pressure, and is expressed in millibars.

Normally air is not completely saturated, so it has an actual vapour pressure which is lower than the saturated vapour pressure. It is useful to compare the actual moisture content to the maximum possible for that temperature and this ratio is the "Relative Humidity" (RH).

$$\text{Relative humidity} = \frac{\text{actual vapour pressure}}{\text{saturation vapour pressure}} \times 100\%$$
$$\text{(at the same temperature)}$$

The saturation factor for a given temperature is expressed in millibars and is obtainable from published hygrometric tables.

Dew Point

If air which is saturated with water vapour is cooled, some of the moisture will be precipitated in the form of mist or dew. Dew point is the temperature at which precipitation begins to appear. Expressed in another way, dew point is the temperature at which the actual vapour pressure becomes equal to the saturated vapour pressure. That is, the temperature when the relative humidity becomes 100%.

Measurement of Moisture Content

The normal method of determining the moisture content of the air is to compare the temperature readings of wet and dry bulb thermometers. The dry bulb temperature is measured by an ordinary mercury in glass thermometer. This is affected only by sensible heat and indicates the air temperature.

The wet bulb temperature is obtained by a similar mercury in glass thermometer, but the bulb is covered by a muslin sleeve which is kept wet by a small reservoir of water. Air is made to pass over the bulb and, if the relative humidity is below 100%, water will be

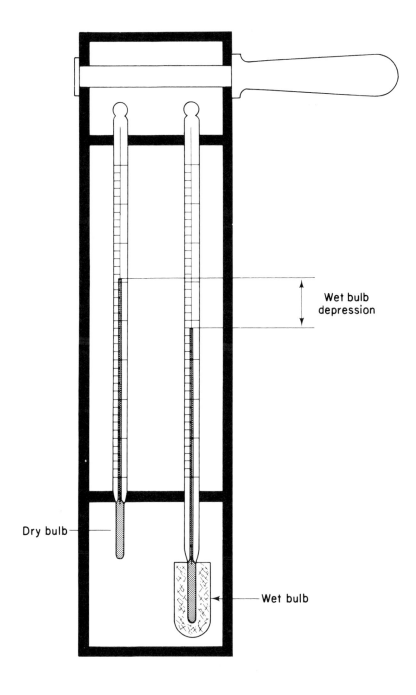

Fig. 1 Sling psychrometer

evaporated from the muslin. The evaporation takes heat from the air and from the bulb of the thermometer which consequently gives a lower reading than the dry bulb thermometer. The difference in the readings is called the "wet bulb depression".

The rate at which water can be evaporated from the muslin around the wet bulb depends on:

- the relative humidity of the air
- air temperature
- velocity of air past the thermometer

Wet and dry bulb temperatures can be accurately obtained by using a "Sling psychrometer", Fig. 1. This consists of a flat frame in which are set similar wet and dry bulb thermometers. The frame is pivoted on the handle so that it can be rotated to move the wet bulb through the air. A small water reservoir is situated at the other end of the frame. The rate at which the instrument is rotated and the minimum time of rotation required for an accurate reading are specified by the manufacturer. Wet and dry bulb psychrometers are used at meteorological stations but there the thermometers are static and enclosed within a "Stevenson Screen". This allows air to circulate around the instrument but gives a shaded, draught free area. There are considerable differences in the wet bulb readings obtained from sling and screen methods and for air conditioning calculations sling temperatures are used.

Hand held electronic instruments are obtainable which give direct digital read-outs.

Psychrometric Charts

Various authorities produce sets of tables and graphs which combine together the information required for calculations relating to the condition of the air. One of the charts in common use is that produced by the C.I.B.S. This psychrometric chart brings together the following information:

- dry bulb temperature
- wet bulb temperature
- moisture content
- relative humidity
- total heat
- specific volume

There are various types of psychrometric chart and a typical psychrometric chart is shown in Fig. 2. This has dry bulb temperature plotted on the horizontal (x) axis and moisture content on the vertical (y) axis. On some charts a series of curves are plotted showing

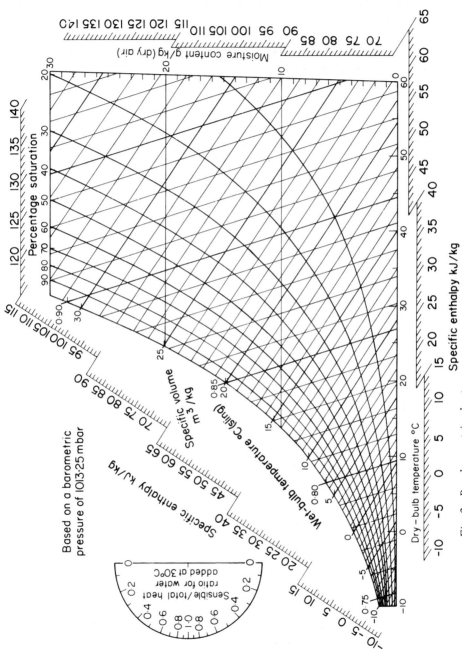

Fig. 2 Psychrometric chart

moisture content against dry bulb temperatures for a range of relative humidities from 10 to 100%. Wet bulb temperatures obtained by the sling psychrometer are plotted along the 100% RH curve, where they are identical to the dry bulb temperature. The wet bulb temperatures appear as diagonal lines across the chart, sloping downwards to the right.

Total heat lines may also be plotted, they run diagonally from top left to bottom right. They are not drawn on the main part of the chart because they are almost, but not quite, parallel to the wet bulb lines. The values may be read by placing a ruler or straight edge across the chart.

Total Heat (Enthalpy)

Total heat is the sum of the sensible and the latent heat contained by a volume of moist air. Sensible heat is indicated by the dry bulb temperature. If more sensible heat was added to the air, its temperature would rise in proportion. Latent heat is the heat required to change the state of a substance. For example, 2250 kJ are required to change 1 kg of water into steam at 100°C. So, if 1 kg of water is evaporated into a volume of air, heat must have been added to the water.

If the heat was drawn from the air, then the air temperature will have been reduced. In effect, some of the sensible heat will have turned into latent heat and the total heat will still be the same as before.

If the heat to vaporise the water is supplied from an outside source, the air temperature will remain constant. But the total heat will have been increased by 2,250 kJ, or the latent heat of vaporisation of the 1 kg of water.

So,

> total heat = sensible heat + latent heat. The total heat energy content of moist air is called "enthalpy" (symbol H) measured in kJ per kg of dry air, above a datum temperature.

Using this concept it is possible to calculate the total heat required to raise the temperature of a given volume of air and to change its RH to the required value by evaporating a quantity of water. In this way the output of the air conditioning plant required may be determined.

The output of a cooling plant is sometimes expressed in "tons of refrigeration" (TR). This is based on the amount of heat absorbed by melting a "ton". This is the "short" or "net" ton, 2000 lb. The "long" or "gross" ton is 2240 lb. of ice at 0°C to water at 0°C in 24 hours. The heat required is 288 000 Btu/day or 12 000 Btu/h which is equivalent to 3.52 kW.

§159   Types of Plant

From the foregoing principles it can be seen that a simple way of reducing the water content of the make-up air is to reduce its temperature below the dewpoint. The surplus water vapour will condense out and give up its latent heat of vaporisation. The heat must be removed from the system by cooling the air, or "chilling". This is done by passing the air over a bank of pipes through which is passing chilled water at about 7°C (45°F).

The equipment used to provide "chilling" is based on the two types of unit used for domestic or commercial refrigerators and described in Volume 2, Chapter 6.
These are:

- absorption units
- vapour compression units

The absorption principle has the advantage that there are no major moving parts in the unit. Small pumps are required to circulate the chilled water and the cooling liquids on some types but the unit is substantially free from vibration and noise.

§160   Absorption Units

There are two main types of absorption unit:

- direct fired, in which gas is burnt to heat the generator or boiler
- indirectly fired, in which the heat is obtained from steam or hot water circulating through coils in the generator, sometimes called a "concentrator"

The direct fired units are of three types as follows:

Direct Fired Air Cooled Chillers

These use ammonia as the refrigerant and water as the absorbent. The condenser and the absorber are cooled by an air current from an integral fan, so avoiding the need for a cooling tower and cooling water.

Because the fan creates some noise and the refrigerant is ammonia, the unit is always mounted externally, usually on a flat roof. Several units can be used on one building to cover large chilling loads, so spreading the weight load on the structure.

Packaged Combination Chillers/Heaters

These use the air cooled chillers previously described, married to gas fired warm air heaters or to steam or hot water heaters supplied from remote boilers. They must be externally mounted and are particularly suitable for large, single storey buildings such as supermarkets. A typical example is shown in Fig. 3, assembled for roof mounting.

Fig. 3   Packaged combination unit

Direct Fired, Water Cooled Chiller/Heaters

The units at present available use water as the refrigerant and lithium bromide as the absorbent. They may be sited anywhere in the building but the condenser and absorber are cooled by circulating water which must be re-cycled through an external cooling tower. This may limit the number of possible locations. When heating is required, the cooling water flow is stopped and the chilled water circuit acts as a heating circuit.

The following table summarises the characteristics of the direct fired chilling units.

§161 **TABLE 1** Direct Fired Absorption Units

Type	Heating Range	Cooling Range	Absorption System	Cooling Tower needed	Siting
Direct Fired Air-cooled chiller		10.5 kW (3 TR) to 35.2 kW (10 TR)	Ammonia/ Water	No	External only
Packaged combination units	26.4 kW (90 000 Btu/h) to 71.4 kW (243 750 Btu/h)	10.5 kW (3 TR) to* 35.2 kW (10 TR)	Ammonia/ Water	No	External only
Chiller/ heater	84.5 kW (288 000 Btu/h) to 141 kW (480 000 Btu/h)	52.8 kW (15 TR) to 88 kW (25 TR)	Water/ Lithium Bromide	Yes	Internal or External

*Up to 106 kW to special order

Indirectly Fired Units

These use water as a refrigerant and lithium bromide as an absorbent and operate under a partial vacuum. Cooling is always by water recirculated through a cooling tower and heat is supplied either by low pressure steam or hot water at temperatures from 104 to 132°C (220 to 270°F). Because the only moving parts are small pumps, usually integral with the machine, they are relatively quiet in operation and may be installed in a wide range of locations. An example is shown in Fig. 4.

The operation of a unit is illustrated in Fig. 5. This is a typical absorption refrigeration system consisting of:

- evaporator
- absorber
- boiler (concentrator)
- condenser

A refrigerant liquid is evaporated by being sprayed at high pressure into a low pressure space. The expansion causes some of the refrigerant to vaporise and in so doing it absorbs heat from a bank of pipes through which is circulating the water to be chilled.

The refrigerant vapour is recovered by being absorbed into the absorbent liquid. Because absorption takes place more readily at a

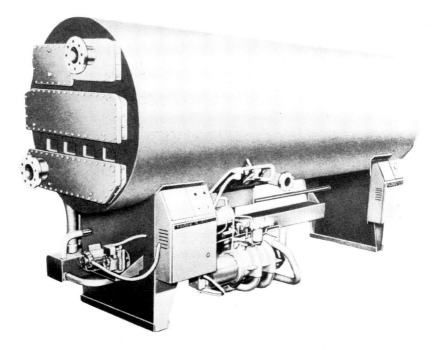

Fig. 4. Indirect fired absorption unit

low temperature, the absorbent is cooled before coming into contact with the refrigerant vapour. The concentrated absorbent liquid having absorbed the refrigerant is pumped back to the generator or concentrator.

In the concentrator, the mixture is heated to boil off the refrigerant as a vapour at a high pressure and the absorbent liquid returns via a heat exchanger to the absorber. The refrigerant vapour passes to the condenser where it is cooled by the coils of cooling water to below its dewpoint. The vapour condenses, forming liquid refrigerant which passes down to the evaporator, so completing the cycle.

§162   Vapour Compression Units

The essential components of the unit in Fig. 6 are:

- evaporator
- compressor
- condenser
- expansion valve

Cooling is brought about by the evaporation of the refrigerant liquid at a reduced pressure in the evaporator. The heat required for vaporisation is absorbed from the coil of pipes through which

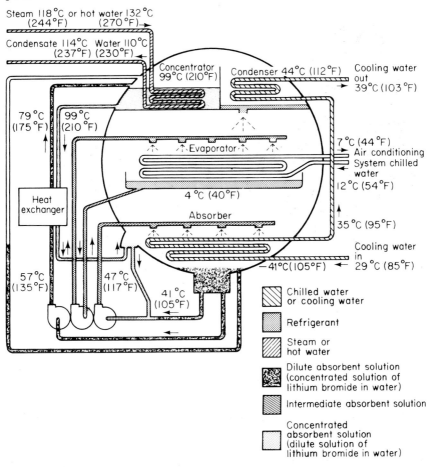

Steam 118°C or hot water 132°C
(244°F) (270°F)→

Condensate 114°C Water 110°C
(237°F) (230°F)

Concentrator 99°C (210°F)
Condenser 44°C (112°F)
Cooling water out 39°C (103°F)

79°C (175°F)   99°C (210°F)

Evaporator

7°C (44°F) Air conditioning System chilled water
12°C (54°F)

Heat exchanger

4°C (40°F)

Absorber

35°C (95°F)

Cooling water in 29°C (85°F)

57°C (135°F)   47°C (117°F)   41°C (105°F)   41°C (105°F)

Chilled water or cooling water

Refrigerant

Steam or hot water

Dilute absorbent solution (concentrated solution of lithium bromide in water)

Intermediate absorbent solution

Concentrated absorbent solution (dilute solution of lithium bromide in water)

Fig. 5   Operation of an absorption unit

circulates the water to be chilled. The refrigerant vapour is removed by the suction of the compressor which compresses the vapour and raises its temperature. The superheated vapour passes to the condenser where it is cooled below its dewpoint and condenses to a liquid. Cooling is effected by a coil of pipes containing water recycled from a cooling tower or by air blown by a fan.

The cooled refrigerant liquid, under pressure, is passed through an expansion valve to the lower pressure side of the system. Here some of the liquid vaporises, further cooling down the remainder. The mixture of cool liquid and vapour passes on to the evaporator, so completing the cycle.

The refrigerants used in compression units include:

· ammonia
· carbon dioxide

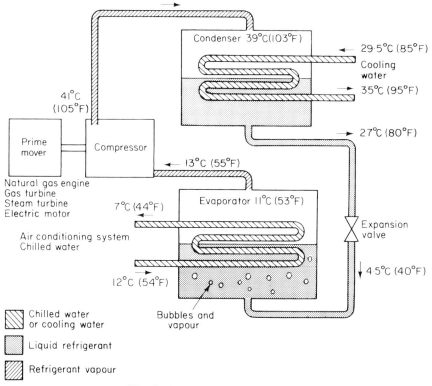

Fig. 6   Operation of a vapour compression unit

- fluorinated hydrocarbons
- methyl chloride

There are three main types of unit classified by the type of compressor used, as follows:

Reciprocating Piston Compressors

These are used for cooling up to 600 kW (170 TR). They can operate at speeds up to 2800 rpm but are usually driven by four-pole synchronous 50 Hz a.c. motors running at 1400 to 1500 rpm or by gas engines. When electrically driven they are constant speed machines and any variation required in the load is achieved by unloading some of the cylinders. When gas engine driven they run at 1750 rpm with load variation being achieved by stepped speed changes and then cylinder unloading.

Centrifugal Compressors

These compressors operate at speeds up to 24,000 rpm.
Single-stage units are available giving 350 to over 3500 kW (100 to 1000 TR).
Multi-stage units give up to 17 500 kW (5000 TR) cooling.

Screw Compressors

These rotate at speeds of about 2900 rpm. They are available in sizes up to 2200 kW (625 TR) cooling.

163 Prime Movers

The "prime mover" which drives the compressor may be one of the following:

- a.c. electric motor
- steam turbine
- gas turbine
- gas engine

a.c. Electric Motors

The normal a.c. electric motor runs at a constant speed, so a compressor must have some cylinders unloaded to meet reductions in load as required. This reduces the operating efficiency of the unit. The electric motor on a unit of American manufacture is designed to operate on a 60 Hz a.c. supply. When used in the U.K. on 50 Hz a.c. it will run at 14% below its design speed. The substitution of an alternative type of prime mover running at 1750 rpm will restore the compressor to its full output capacity.

Steam Turbines

These can operate at the speeds required by centrifugal compressors and their speed can be varied. They are generally available in only a small number of large sizes.

Gas Turbines

These have similar characteristics to steam turbines but eliminate the need for a boiler.

Gas Engines

Generally these are modified versions of standard piston engines originally developed for diesel oil or petrol. They are available in sizes from about 19 kW (25 hp) upwards.

164 Gas Engines

There are two main types of gas engine:

- dual fuel (compression ignition)
- spark ignition

Dual Fuel Engines

Similar to diesel engines, these can be run either on oil or on a mixture of gas and "pilot" oil. The air/gas mixture is compressed in the cylinders and is ignited by the injection of a small amount of oil. The injected "pilot" oil is usually between 5 and 10% of the full fuel requirement of the engine.

Engines are available in sizes from 134 kW (180 hp) up to several thousand kW. They are heavy for their output and require frequent and highly skilled maintenance. In consequence they are used for large installations and are not generally marketed as packaged units with compressors.

Spark Ignition Engines

In these engines the air/gas mixture is ignited by a high tension electric spark as in a petrol engine. They are lighter and easier to maintain than dual fuel engines and their initial and installation cost is less. The types available may be either industrial engines or converted automotive engines. Industrial engines are available in sizes from 37 to 1566 kW (50 to 2100 hp) with running speeds ranging from 1200 to 2400 rpm and compression ratios of up to 10:1. Converted automotive engines are simply ordinary car or lorry engines modified to run on natural gas instead of petrol. A range of converted engines is available and these can be obtained linked to chilling compressors of 70 to 5275 kW (20 to 1500 TR) output.

An example is shown in Fig. 7.

Fig. 7   Reciprocating compressor, dual service set

165 Installation of Gas Engines

Mounting

A heavy dual fuel engine must be mounted on a separate base which is isolated from the building structure so that vibration is not transferred. Spark ignition engines are generally light enough to be coupled to the compressor and the whole unit mounted on a steel frame or "skid". The skid is then set on anti-vibration mountings as in Fig. 7. Large spark ignition engines are mounted on mass concrete bases which are considerably cheaper than an equivalent steel base. There is generally less trouble with vibration from spark ignition engines because of their lighter weight and higher speed.

Noise

The compressor of a chilling unit is sufficiently noisy to require careful siting and the substitution of a gas engine for an electric motor makes it essential that steps are taken to reduce the noise level.
The additional engine noise comes from:

- mechanical noise from the engine
- the cooling equipment
- the exhaust system

Acoustic cladding of the plant room walls and possibly hoods over the engine and compressor can reduce the effect of mechanical noise. Fan and radiator cooling systems tend to be noisier than water to water heat exchangers. Exhaust noise is reduced by suitable silencers and by siting the discharge away from any sensitive areas.

The unit should be connected to its outlet ducting by flexible joints to avoid the transmission of noise from the plant.

Air Requirements

Air is required for combustion and ventilation. Combustion air is required for the engine and any other fuel burning equipment. For the engine this is about 10 times the gas flow rate.

For ventilation the air requirement depends on:

- the permitted air temperature rise in the room
- the method of cooling the engine water jacket
- the degree of insulation of radiating surfaces, in particular the exhaust manifold and system.

In each case the plant manufacturers should be consulted and any other equipment in the room also taken into consideration.

A major engine manufacturer recommends the following formula:

$$m^3/s = \frac{0.1406 \times kW}{t°C} \quad \text{or} \quad ft^3/m = \frac{400 \times hp}{t°F}$$

where:

m^3/s = cubic meters/second $\}$ air required for ventilation
ft^3/m = cubic ft/minute
kW = rated power in kilowatts $\}$ of the engine
hp = rated horse power
$t°$ = permitted rise in air temperature

Engine Cooling

Engines are water-cooled. Water circulates through passages in the cylinder block and a difference in temperature of 10 deg C (18 deg F) should be maintained between the inlet water at 70 to 80°C (158 to 176°F) and the outlet at 80 to 90°C (176 to 194°F). Actual temperatures vary with different makes of engine. Thermostats are used to control the flow of cooling water and a high temperature cut-out is often included.

The hot water leaving the engine is cooled by means of:

· water to air heat exchanger (radiator)
· water to water heat exchanger

Radiators with fans are simple but noisy and are generally used for smaller engines.

Water cooling of the engine water requires a cooling tower but a tower is already required for the refrigeration condenser and one cooling tower can serve both units. The temperature of the cooling water leaving the condenser is low enough to be used for the inlet water to the heat exchanger and only one circuit is required.

Exhaust

The temperature of the exhaust gases depends on:

· type of engine
· fuel used
· engine loading

Turbo-charged engines exhaust temperature is typically 420 to 460°C (788 to 860°F). Other engines are generally 550°C (1022°F) or below and a few may reach 650°C (1202°F). The hot gas must be conveyed safely to discharge outside the building in accordance with the recommendations of the manufacturer.

The following points must be considered:

- the position at which the exhaust terminates
  - –clear of air inlets to the building
  - –unlikely to cause noise nuisance
  - –clear of adjacent buildings
  - –in accordance with planning requirements
- exhaust pipe diameter should be at least as large as the manifold connection.
- total backpressure of piping and silencers to conform to makers specification
- exhaust piping to be capable of withstanding maximum exhaust temperatures
- a high-temperature flexible connection to be used between the engine exhaust manifold and the exhaust pipe
- the piping must be protected by insulation or ventilated sleeves so that it does not present a fire hazard inside the building
- precautions should be taken to avoid condensation; where moisture is likely to be trapped, drainage points must be provided
- provision must be made to accommodate the thermal expansion of the exhaust piping
- adequate support is necessary to avoid damage to the flexible connection.

Engine Starting

Large dual fuel engines are usually run up to starting speed by injecting compressed air. Smaller engines and spark ignition engines are usually started by means of electric starter motors powered by lead/acid batteries. The batteries are trickle charged from the mains supply. They require only a weekly check on acid level and an occasional check on specific gravity.

Some spark ignition engines have a compressed air motor mounted in place of the electric starter.

Gas Supply

The gas meter and the gas supply to the engine must be capable of meeting the maximum consumption demanded by the engine on full load, without reducing the pressure to any other appliances on the same supply.

Industrial engines must be fitted with a low pressure cut off switch in the supply, as close to the engine as possible but upstream of the final governor.

Engines with natural induction require gas at pressures below 200 mbar.

Turbo-charged engines require gas at pressures of about 2 bar. Usually the pressure is supplied by a compressor.

Warning notices should be fitted at the engine starter and at the meter inlet, Chapter 2, section 33.

§166 Dual Service Sets

Supermarkets and some other commercial undertakings require emergency lighting at about 40% of the normal load in case of electric supply failure.

When a gas engine is used to drive a compressor, it can also be coupled to an alternator. Figure 7 shows a dual service unit giving up to 60 kVA in which the alternator is driven by three Vee belts and the coupling to the compressor is through an electric clutch.

When the electric supply fails, the electric clutch disengages the compressor, the engine speed is automatically adjusted to 1500 rpm and current is available in the stand-by circuits at 50 Hz a.c. A reasonable level of emergency lighting is provided at the expense of the chilled water. However, the cooling load is reduced because of the reduction in lighting. If the engine is stationary at the time when the supply fails, it is automatically started, declutched and run up to generating speed.

§167 Controls

Controls are fitted to the sets to prevent overheating, ensure safe operation and obviate any false starts. These may include:

- oil pressure switch, to shut down the engine if the oil pressure falls too low
- time delay switch, to prevent too rapid cycling by stopping the compressor for about 5 minutes after each shut-down
- low pressure switch, to stop the compressor if suction becomes so low that it would produce freezing in the chilled water
- high pressure switch, to shut down if pressure rises excessively for any reason

Most plants have safety switches fitted in the chilled water circulation to stop cooling if the water temperature approaches freezing point.

168　Plant Servicing

Much of the maintenance work on air conditioning plant is carried out by specialists in this field. Organisations which have large installations often have a resident plant engineer. The plant engineer carries out daily and other periodic checks and operations. Annual servicing is normally done by a specialist firm of refrigeration engineers.

Water Chillers

Servicing of gas fired chillers and chiller/heaters should include:

- cleaning combustion chamber and flueways
- checking and adjusting gas rate
- checking operation of gas and electrical controls
  - —overheat switches
  - —thermostats
  - —safety cut-offs

Some machines require non-condensible gases to be purged and on air-cooled chillers the air ways must be cleaned and the air flow checked. Belt drives should be examined and belt tension checked.

Gas Engines

Major services are carried out on the basis of the number of hours of running. Most engines are fitted with an "hour-run" meter.

Oil and water levels should be checked daily and battery acid levels every week. Spark plugs, points, oil and filters should be changed periodically as specified by the manufacturer. The requirements for servicing are less than those for a comparable petrol engine.

Servicing schedules are provided by most manufacturers and some offer maintenance contracts.

169　Air Curtains

Many air conditioned buildings have two sets of outside doors forming an air lock to prevent any sudden in-rush of un-conditioned air. Where only single doors are used, the incoming air may be dissipated by a stream of air directed downwards or across the opening. Some air curtains use only cold air but more frequently, in this country, warm air curtains are employed. The warm air is produced by a directly or indirectly fired unit or by a steam or water to air heat exchanger.

Fig. 8   Air curtain, door heater

A typical gas fired heater for large doors is shown in Fig. 8. This is a direct fired air heater, with the burner situated in the path of the air flow through the heater. Twin centrifugal fans expel the hot air downwards through the outlet louvres. The heat input can be varied from 102 kW (350 000 Btu/j) upwards to suit a particular doorway. Mounting heights are generally between 3.5 and 7.5 m (10 to 25 ft) and the air inlet is located outside the building, unless the operation of the heater is interlinked with the door opening and closing. Because the installation of a direct fired heater contravenes the Building Regulations, a waiver must be obtained before it may be installed.

The British Standards Institution recently issued a revised Code of Practice, BS 5720 (1979), "Mechanical ventilation and air conditioning in buildings", which deals with the work involved in design, installation, commissioning, operation and maintenance of mechanical ventilation and air conditioning systems.

CHAPTER 11

Large Scale Heating and Hot Water Systems

Chapter 11 is based on an original draft by Mr. B. J. Whitehead

170 Introduction

Whilst hot water systems are heated by various designs of gas fired boilers, heating systems can utilise either boilers or warm air heaters. The choice of system depends on a variety of factors including:

- whether building is existing or projected
- type and construction of building
- size and purpose of building
- frequency of occupation
- environmental requirements of activities or processes carried out within each room
- limitations imposed on projected capital and running costs

In many cases buildings or parts of buildings may be used for very different activities, each requiring its own level of heating. For example, in hotels and some public buildings a room used for a conference during the day may be used for a dance in the evening and the system must be flexible enough to provide for both. In buildings where not all of the rooms are in continuous use a zoned system or individual independently controlled emitters may be required. Libraries, museums and art galleries must maintain constant levels of temperature and humidity and electrostatic filters and humidifiers may be used. Many of the supermarkets and departmental stores have systems which need to exhaust vitiated air and introduce a quantity of fresh air to maintain a pleasant environment for their customers, however 'total loss' systems are now outdated, in view of the energy situation.

§171   Warm Air Heating

Warm air heating units are available in four main forms:

- unit heaters—for overhead suspension with or without ducting systems
- horizontal and inverted heaters—for overhead installation
- direct fired air heaters (e.g. make up air heaters)
- floor standing heaters—with or without ducting systems

All types are generally suitable for installation in existing and projected buildings and have the added advantage of providing fresh air circulation or air re-circulation at ambient temperature during warm weather. Units may also be fitted with filters to clean fresh or re-circulated air. This is desirable in shops, restaurants and public houses.

The first three types of heaters were introduced in a previous chapter.

§172   Floor Standing Heaters

Construction

The units are designed to draw in cool air at floor level and to discharge heated air through one or more outlets at the top of the heater, Fig. 1. The outlets may take the form of fixed or swivel directional nozzles or a connection to a duct system.

Most heaters are indirectly fired and the heat exchanger is usually tubular, made of stainless steel and may have baffles in the tubes to improve heat transfer. The air is forced through the heater by one or two large capacity centrifugal fans situated in the lower section. The fan motor is often 380/440V a.c. 3 phase and may range from 1.0 kW (1.3 hp) up to about 11.0 kW (14.8 hp).

Automatic forced draught burners are used in conjunction with full sequence control and flame rectification or ultra violet flame monitoring. Other controls may include:

- air pressure switches; activated by combustion air or flue fan pressures
- limit stat
- fan delay unit
- thermostats; appliance, day and night or night set back types
- clock

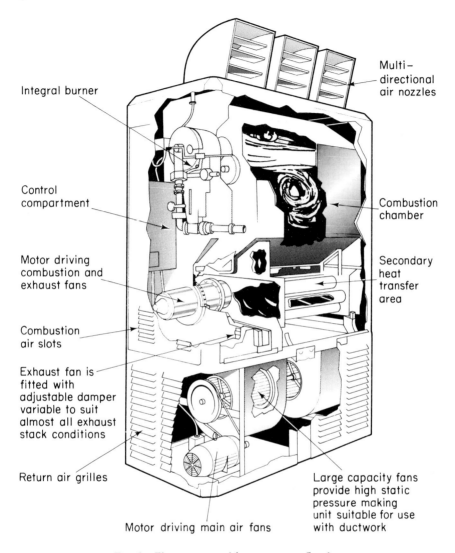

Integral burner

Multi-
directional
air nozzles

Control
compartment

Combustion
chamber

Motor driving
combustion and
exhaust fans

Secondary
heat
transfer
area

Combustion
air slots

Exhaust fan is
fitted with
adjustable damper
variable to suit
almost all exhaust
stack conditions

Return air grilles

Large capacity fans
provide high static
pressure making
unit suitable for use
with ductwork

Motor driving main air fans

Fig. 1   Floor mounted heater, open flued

Heater outputs range from about 44 kW (150 000 Btu/h) up to 1000 kW (3 500 000 Btu/h). Gas connections are usually Rc1 to Rc1$\frac{1}{2}$ and flues from 200 to 250 mm (8 to 10 in) diameter. Heaters are open flued which may have fan assisted or fan diluted systems added.

§173   Warm Air Systems

Free Discharge Systems

When designing a heating system based on direct discharge from a floor mounted heater, the position of the unit is of paramount importance. The following points should be noted.

Warm air must be free to discharge evenly over the area to be heated without obstruction. Manufacturer's data normally gives information on the effective throw of the heated air output. This may be from 15 to 45 m (50–150 ft).

A clear space must be left around the base of the heater to allow the return of cool air for recirculation. When return air is taken in on all sides a distance of 450 mm (18 in) should be allowed between the heater and any obstruction. Recirculation intakes must not be restricted or blanked off without prior agreement with the heater manufacturer.

The movement of return air to a heater may be noticeable within a radius of about 2 m (6 ft) depending on heater size. The heater should be sited so that no permanent places of work are within this area.

Typical locations for effective warm air distribution are shown in Fig. 2.

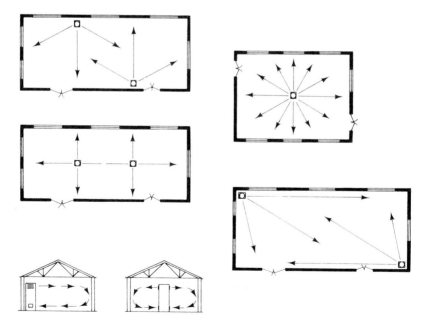

Fig. 2   Location of freestanding floor heaters

Ducted Systems

Where considerations of building layout, furniture and fittings, appearance or noise factors make it impracticable to fit the heater in the working area a duct system can be used. Systems may be designed to distribute heat evenly throughout a large area or to convey varying quantities of warm air into separate rooms wherever practicable diffusers should be sited or designed so that the air flow does not come into contact with the occupants of the room. Duct-work is more easily installed in buildings during their construction. It may, however, be fitted in existing buildings particularly where it can be hung openly under ceilings, or concealed in false ceilings or under suspended floors. In certain situations fire dampers or smoke detection equipment may be necessary under Local Authority by-laws or Fire regulations. Examples of various duct systems are given in Figs. 3 to 7.

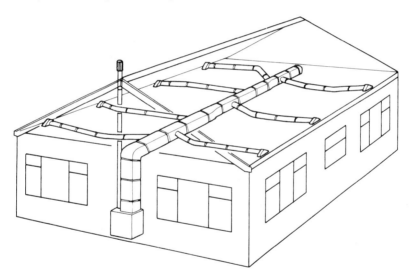

Fig. 3   Simple overhead duct system

Fig. 3 shows a simple overhead duct system from a floor mounted heater. Warm air is delivered by ceiling diffusers supplied by flexible circular ducting from a stepped duct system. Air returns via the grilles on the heater unit. All ducting in the roof space is insulated with 50 mm (2 in) of glass fibre or similar material. It would be possible to save floor space by using a horizontal heater mounted overhead.

Another application of an overhead duct system to a single storey building is shown in Fig. 4. This is a shop where the heater is fitted

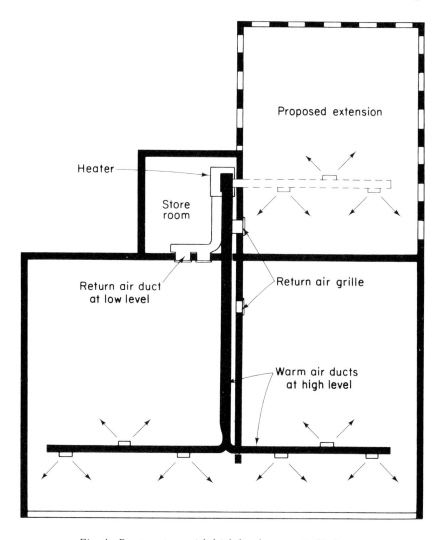

Fig. 4   Duct system with high level warm air discharge

unobtrusively in a store room at the back. Ducts are run under the
ceiling with high level outlets. As in the previous example, this
requires air velocities of about 3.5 to 5 m/s (690 to 990 ft/min) to
take the air down to floor level. The high velocity creates noise
which, in some cases, may be unacceptable but would probably be
unnoticed against the background noise in a town centre. It is possible
at the design stage to calculate the noise emission factor.

Figure 5 shows a single level duct system serving two storeys in a
building. The heater is fitted in the basement and the ducting run on

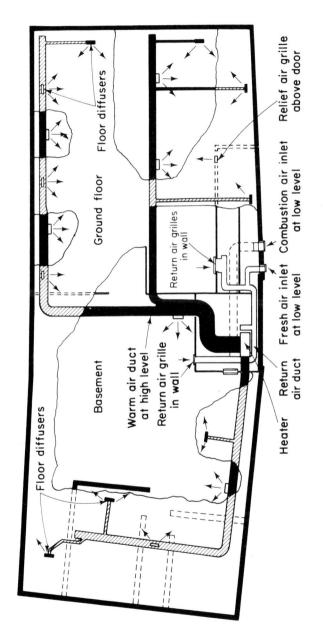

Fig. 5 Duct system in 2-storey premises

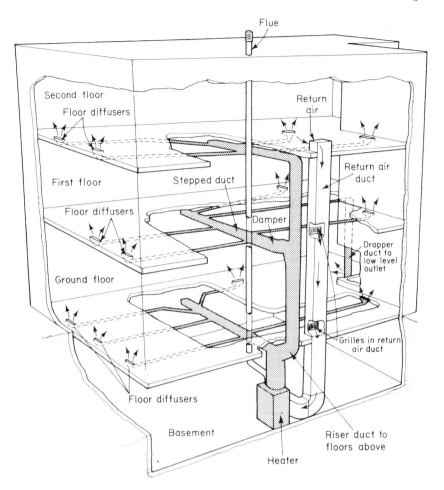

Fig. 6   Three-storey building with ducts in suspended ceilings

the basement ceiling. Ceiling level outlets serve the basement and the ground floor is heated by floor diffusers.

Larger buildings with several floors may be heated by the type of system in Fig. 6. The ducting is concealed above suspended ceilings using floor diffusers to supply the floor above. The return air is ducted from each floor by grilles at low level in the return air duct.

Where a building is composed of various sized rooms with widely different heat requirements, the use of two or more heaters should be considered. This allows simple duct systems to be used and gives flexible operation of the whole system. Fig. 7 shows three heaters

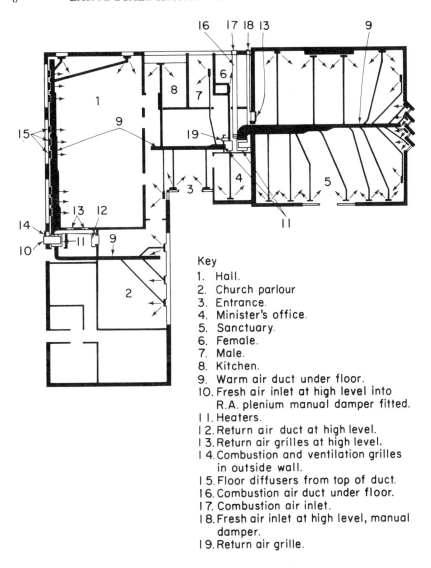

Key
1. Hall.
2. Church parlour
3. Entrance.
4. Minister's office.
5. Sanctuary.
6. Female.
7. Male.
8. Kitchen.
9. Warm air duct under floor.
10. Fresh air inlet at high level into R.A. plenium manual damper fitted.
11. Heaters.
12. Return air duct at high level.
13. Return air grilles at high level.
14. Combustion and ventilation grilles in outside wall.
15. Floor diffusers from top of duct.
16. Combustion air duct under floor.
17. Combustion air inlet.
18. Fresh air inlet at high level, manual damper.
19. Return air grille.

Fig. 7   Church with three independent heaters

installed in a church and operating independently according to building use. The heater serving the sanctuary draws its combustion air from outside via a duct below the floor. A fresh air inlet is also provided to feed into the high level return air duct. 10 to 15% of the air returned is fresh air.

§174　Installation of Floor Mounted Units

Location

In addition to any space required for return air, access must be provided for servicing or replacement of the equipment. Manufacturer's instructions should be consulted. Where the heater is in contact with combustible materials, suitable heat insulating shields should be provided in accordance with the maker's specification.

Air Inlets and Discharge Outlets

The heater depends for efficient operation on the free flow of air through the heat exchanger. Any restriction of inlets or outlets can adversely affect the operation. Precautions must be taken to prevent obstruction occurring and guard rails may need to be provided.

In addition, where there is a risk of combustible material being placed near to the warm air outlets, suitable barrier rails should be fitted to maintain a distance of 1 m (3 ft) between the material and the outlet.

The minimum distance between the heater inlet and any obstruction should be 450 mm for 117 kW (18 in for 400 000 Btu/h) rising in proportion to the heater output.

In special circumstances where air may be polluted by dust or shavings, the air inlets should be positioned to avoid contamination and screens may be necessary to trap the pollution.

When return air inlets are close to extraction systems carrying flammable vapours, precautions must be taken to prevent any vapour being entrained with the return air. This may entail ducting the return air from a safe distance. In such cases system design is even more critical and compliance with the additional safety requirements essential.

The installation of heaters in buildings where there is a fire hazard from flammable vapours should conform to the requirements of the local petroleum or fire regulations. As a general rule, combustion air should be ducted to the heater from outside the building. The duct must be of adequate size and its inlet should be positioned clear of any low pressure zones, areas of wind turbulence, and high enough above ground level to prevent accidental blockage from snow, dead leaves, etc.

Prevention of Noise Transmission

Where noise level is important, the heater should be insulated from the structure of the building by standing it on felt pads or anti-vibration mountings. When this is done it is essential that all fuel, ducting, flue and electrical connections are made flexible. Otherwise vibrations may still be transmitted and the connections may be damaged.

Noise transmission into ducting can be reduced by using a non-flammable, sound insulating connection between the heater and the duct run. Where a heater is installed in a separate compartment, it may be necessary to fit sound "attenuators" or reducers in the duct run before it enters a sensitive area. Manufacturers can advise on this aspect.

Heater Compartments

To comply with architects requirements or local regulations or to cater for those installations where fire hazard, noise or risk of tampering must be minimised, it may be expedient to site the heater in a separate compartment. Requirements for compartments are broadly similar to those described in Volume 2. The main points to be considered are as follows:

The compartment must be constructed of stable, fire resistant materials. It may be acoustically lined to minimise noise.

There must be adequate access available only to authorised persons for servicing, inspecting and renewing the equipment.

Provision must be made for:

- a supply of air for combustion
- circulation of cooling air to remove heat emitted from the flue and any duct work in the compartment
- entry of return air

Any air inlets to a compartment should be clear of sources of smells or fumes. If noise may be emitted from an air inlet, the inlet should be positioned to cause the minimum of nuisance.

Flues

Flues should be supported independently of the heater unless otherwise specified by the manufacturer.

Precautions must be taken to prevent rain water or condensation from the flue from entering the heater. A condensation trap should be provided at the base of the flue.

§175   Servicing

Servicing should be carried out annually or after every 1000 hours operation whichever is the shorter.

The servicing of large warm air heaters follows a similar procedure to that specified in Volume 2, Chapter 11 for domestic models. The major differences are that the larger heaters are normally fitted with:

- 415V a.c. 3 phase fan motors
- automatic forced draught burners
- flame rectification or ultra violet flame sensors

Methods of checking and servicing these components have been dealt with in previous chapters.

The remaining components which require special attention are the heat exchanger and the filter, if fitted.

The heat exchanger and the combustion chamber should be examined for corrosion or blockage, using an inspection lamp. Follow maker's instructions on methods of gaining access. Clean out the heat exchanger tubes using a scraper or a brush and remove deposits with an industrial vacuum cleaner.

The normal atmosphere carries a considerable quantity of dust and filters can quickly become choked if not regularly cleaned. A regular cleaning schedule should be established.

§176   Central Heating by Hot Water Boilers

The introduction of natural gas has been followed by new developments in commercial boiler design and in boiler systems. Many of the older designs of boilers are still to be found on the district and the principal types to be met are:

- cast iron sectional boilers
- steel shell boilers
- modular boilers

Cast Iron Sectional Boilers

These are built up of a number of cast iron sections which contain the waterways and form the combustion chamber, Fig. 8. The sections are joined by nipples to form a flow header at the top and left and right return headers at the bottom.

Boilers are made up of the number of sections required to give the desired output capacity. They may be fitted with jetted bar natural draught burners or with automatic forced draught burners.

Both types of burner incorporate some form of automatic control and flame protection device.

Steel Shell Boilers

These boilers are designed for higher pressures and greater heating capacities than cast iron sectional boilers. They normally operate at higher efficiencies. Shell boilers are generally used where heat outputs of above 586 kW (2 million Btu/h) are required. Models for

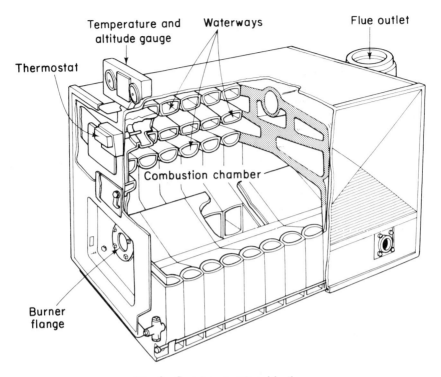

Fig. 8   Cast iron sectional boiler

high pressure hot water and steam are available with heat outputs up to 5860 kW (20 million Btu/h).

Although output ratings are high, the boilers are about 50% smaller in cross-sectional area than conventional sectional boilers of the same output rating.

Some types of shell boilers were described in Chapter 8. Another design is shown in Fig. 9. This consists of two concentric shells enclosing an annular water space through which fire tubes are fitted. The combustion chamber is in the centre of the inner shell and the hot flue gases recirculate through the fire tubes, so giving a high thermal efficiency.

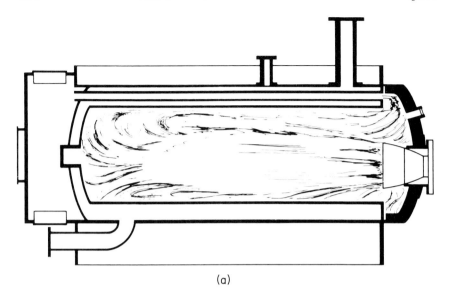

(a)

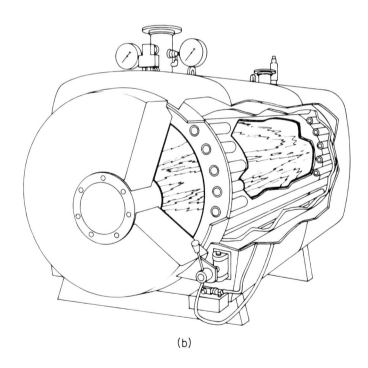

(b)

Fig. 9   Steel shell boiler

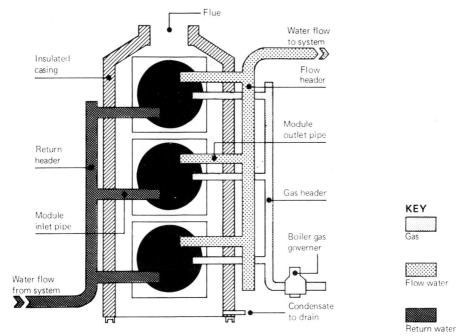

Schematic diagram of the three module Modecon 150.

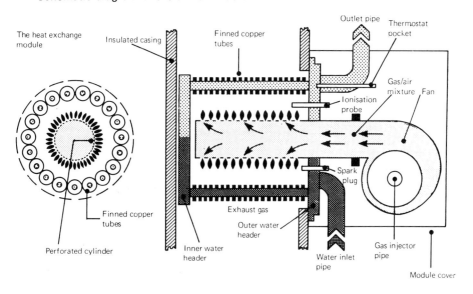

Schematic diagram of heat exchanger and associated controls.

Fig. 10   Modular boiler

Modular Boilers

There are various designs of modular boilers all of which generally possess similar characteristics as follows:

- low thermal capacity, quick response
- light in weight, suitable for roof top installations
- small size, to pass easily through doorways or roof traps and be transported in passenger lifts.
- low noise level
- high thermal efficiency, low running costs
- easy to install and service

Because a modular system generally consists of from two to about six boilers, there is relatively little loss of heat output when one boiler is shut down for servicing or when a breakdown occurs.

A number of modular boilers have natural draught burner systems and are fitted with down draught diverters. Comparatively recent designs have a premix burner and are connected to an open flue system but without a draught diverter. Schematic diagrams of a three module unit and an individual module are shown in Fig. 10.

The heat exchanger consists of finned copper tubes arranged in a circle around the perforated, cylindrical burner. Water enters at the front and passes through the lower tubes to the rear header, returning through the upper tubes to the flow.

The air/gas mixture is controlled by a zero governor which is impulsed to respond to fan pressure and to pressure in the flue chamber. So the governor will shut down if the fan fails or if the flue becomes blocked.

Because the thermal efficiency is remarkably high (85.0%), some condensation will occur in the heat exchangers when starting from cold and in the flue at that and other times. Provision must be made for condensate to flow to a removal point where it can be carried by a 22 mm pipe to a gulley. The flue should be insulated or twin-walled.

The modulus will be shut down in sequence, normally starting at the top. Control may be a step sequence controller or by boiler thermostats in each module. In this way, water passing through the upper modules is heated by flue gases from the lower.

The boiler must be fitted with a pumped circulation and the head loss is about 75 mbar (30 in w.g.) at a temperature rise of 11 deg C (20 deg F).

§177   Emitters

Various types of emitter are available. Some provide heating principally by radiation, others principally by conduction. Most of these
have been described before in previous chapters and it remains only
to summarise the types and add a little further information.

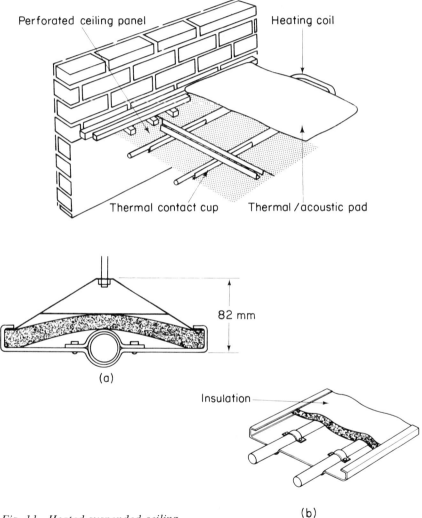

Fig. 11   Heated suspended ceiling

Radiant Heating

This may be provided by:

- heated suspended ceilings
- radiant panels
- radiant strips

These were introduced in chapter 9.

Heated suspended ceilings. A heating coil of comparatively small water content lies in contact cups on perforated ceiling panels of highly conductive metal. Pads of insulating material direct the heat downwards and also act as an acoustic insulator, Fig. 11. The system is slow to respond to changes in heating demand.

Radiant panels. These consist of a pipe coil fixed to a metal plate, usually of steel. There are many designs to suit different applications.

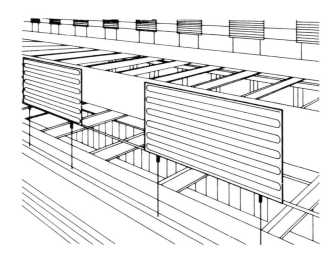

Fig. 12   Radiant panels

They are usually suspended from roof trusses, as in Fig. 12, or fitted to walls at high level. Because they give localised heating they are suitable for entrance canopies, loading bays, workshops and exhibition halls.

Radiant strip. This is similar to the radiant panel but the pipe and the aluminium reflector form a long continuous strip suitable for heating corridors or large spaces where long uninterrupted runs are possible. The units may include lighting fittings and are usually heated by medium or high pressure hot water. The higher surface temperatures produced reduce the length of strip required to about half that required with low pressure water. Fixing heights are given in Table 1.

TABLE 1   Fixing Heights for Radiant Strip Heating

Hot water temperature		Minimum height of strip above floor level			
		One pipe		Two pipe	
°C	°F	m	ft	m	ft
70	160	2.4	8	2.7	9
95	200	2.7	9	3.0	10
115	240	3.0	10	3.3	11
140	280	3.0	10	3.6	12
160	320	3.3	11	4.0	13
180	360	3.6	12	4.2	14

Convector Heaters

Radiators. The older types of cast iron column radiators have greater heating surfaces than the modern steel panels but are heavier and less attractive. They are still used where appearance is not the main consideration, principally in institutional premises. A cast iron

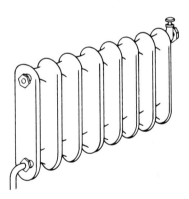

Fig. 13   Hospital radiator

hospital radiator is shown in Fig. 13. More recently it has been possible to get column radiators in cast aluminium, which are quite attractive but more expensive.

Convectors. Natural convectors consist of a water to air heat exchanger enclosed in a steel casing having louvres or grilles at the top and bottom. The output of warm air is usually controlled by a damper at the top of the unit, Fig. 14.

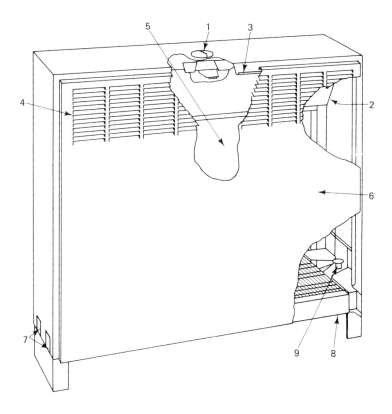

1	Damper control knob	6	Removable front panel
2	Damper blade	7	Knock-outs for connections (both ends)
3	Front panel support channel	8	Heating element
4	Outlet grilles	9	Element support and vertical adjustment
5	Insulated rear panel		

Fig. 14   Convector heater

Wall strip heaters. These are a continuous convector heater designed as a skirting heater or with additional heating tubes in heights up to about 1 m (36 in) for under window heating in offices, Fig. 15.

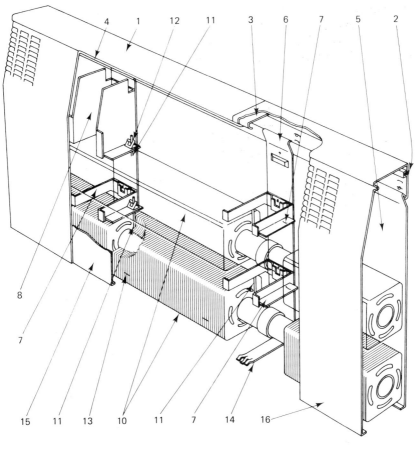

1	Top rail	not shown 9	Frontplate centre support
2	Plastic foam strip		10 Element
3	Inner joint rail		11 Element support hook
4	Top rail stiffener		12 Spire clip
5	Backplate		13 Noise suppression clip
6	Hanging strip		14 Retaining clip
7	Element support bracket		15 Make-up plate
8	Make-up plate/element support bracket		16 Front plate

Fig. 15   Wall strip heater

Fanned convectors. Fan convectors are available with variable speed fans, usually low, normal and booster. Control is by thermostat which switches the fan on or off. Normally when sizing the heater, only output at low and normal settings are used. The use of such equipment needs a little thought in view of the noise which may be

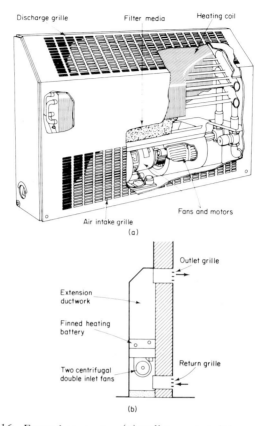

Fig. 16   Fanned convector (a) wall mounted (b) recessed

given off. The heaters are very compact and have a high heat output, Fig. 16(a). They may be fitted in recesses, Fig. 16(b) and are widely used in schools, offices, gymnasia and entrance halls. Sizes vary from 1 to 1.5 m (3 to 5 ft) in width and 1 to 2.5 m (3 to 8 ft) in height.

Heater Batteries

These consist of heating coils of copper or steel finned tubes together with fans and filter. They may be housed in a framework incorporated into a duct system or form a complete heating unit as in Fig. 17.

Heater batteries may be used for:

- warm air heating
- air curtains
- make-up air heating
- air conditioning plants

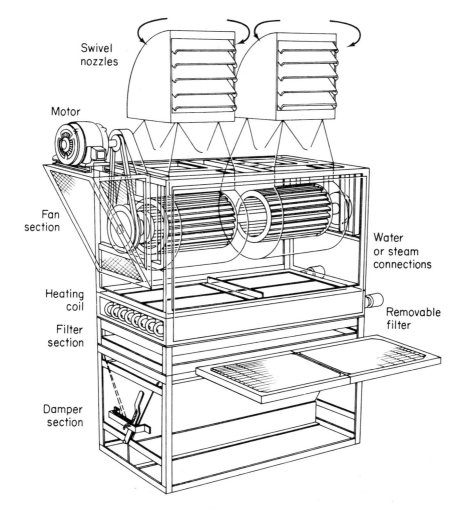

Fig. 17   Heater batteries

They may also be used in conjunction with radiators. In this case the warm air makes good the heat loss due to air change and the radiators compensate for the heat loss through the structure.

Unit Heaters

Unit heaters are small water to air heat exchangers with fanned convection. They are mounted overhead or at high level on walls. Most frequently used in factories or warehouses they are obtainable with horizontal or down flow discharge, Fig. 18.

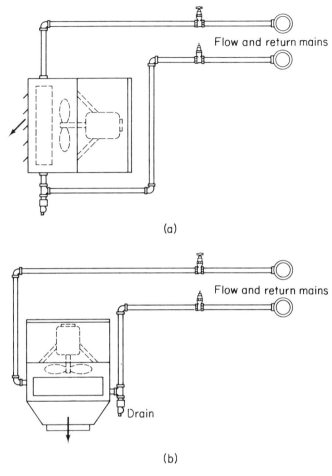

(a)

(b)

Fig. 18   Unit heaters
(a) horizontal flow
(b) down flow

§178   Control Systems

Temperature Compensation

Whilst most domestic central heating systems require only simple control systems, the energy saving possible from a large commercial installation justifies the use of sophisticated equipment. Weather conditions vary considerably and plant is usually designed to give inside temperatures of 18 to 22°C (64 to 72°F) with an outside temperature of about −1°C (30°F). However, the mean winter temperature in the south of England is about 12 to 13°C (53 to 55°F). This gives a load factor on the system of less than 60%. If excess heating is to be avoided, and energy conserved the boiler output must be related to the prevailing weather.

The second problem is that of regulating the heat output to meet day to day demands. This is normally done by splitting a large installation into a number of parts or zones, each under separate control. The pipework must be designed or modified to make this possible. When existing properties are being modernised it is often impossible to split the pipework into zones. For this reason the use of an outside temperature compensator for overall control became popular. By mixing flow and return water in a three port valve the heat input could be regulated in sympathy with the prevailing weather conditions, usually measured on the north wall of the building. Some compensators respond to solar gain and wind velocity as well as temperature. A programmed time switch may be incorporated to give day and night control and early morning boost.

The outside temperature compensator controls the heating circuit independently of the hot water demand when both are fed from the one boiler. The hot water cylinder may be controlled by:

- cylinder thermostat controlling two port valve
- cylinder thermostat controlling separate pump on hot water circulation

An alternative control system to the outside temperature compensator worth mentioning is known as an Optimiser. The principle difference is that an Optimiser also decides the time that the heating installation should be brought on, on a day to day basis, that is if it takes into account the weather prevailing and the type of building in use.

Zone Control

When pipework has been suitably arranged to provide separate zones, their control may be effected by:

- restricting the flow of water to the heating surfaces by two port valves
- modulating the zone heating surface temperature by means of mixed water using three port valves

When restricting the water flow by room thermostats and two port valves, on/off or two position valves are usually suitable.

When heating is by blown warm air it is usually essential to use a modulating valve to avoid a blow hot/blow cold effect on the occupants.

In order to stabilise the pressure drop presented to the circulating pumps, three port diverter valves are sometimes used instead of two port valves. The heated flow water goes either to the heat exchanger or back to the common return but this increases system heat losses at times of low demand.

If the temperature of the heating surface is to be modulated, each mixing loop must either have its own pump or two port valve control.

Figure 19 is a schematic diagram of a building with a long East/West axis. The whole building is weather compensated from the north aspect. The Southern faces are made into separate zones. In the example shown, the South East zone is controlled by a three port valve which allows some of the flow water to bypass the heat exchangers. The South West zone has a two port on/off valve which controls the flow through the zone's heat exchangers.

Restricting the flow to a number of heat exchangers provides a variable volume at constant temperature. When the flow is restricted

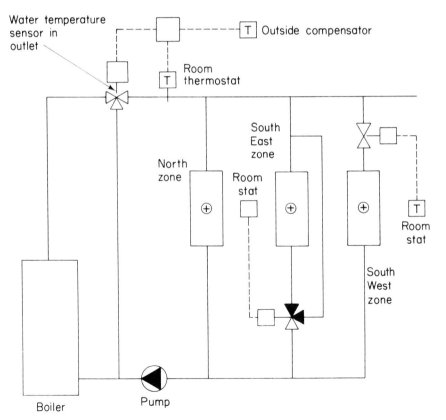

Fig. 19 Zone control

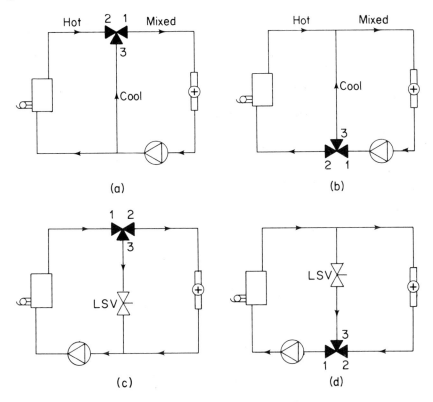

Fig. 20   Use of three-port valve
(a) mixer in flow
(b) diverter in return
(c) diverter in flow
(d) mixer in return

the temperature drop across the system is increased. This can result in the emitters nearest to the source of heat always being hot, even under low load conditions. Those progressively nearer to the system return will be increasingly cooler. The emitters must be grouped in such a way that this effect is minimised, i.e. in parallel not in series.

A mixed water scheme gives constant volume at variable flow temperature and is preferable for large series/parallel groups of emitters, particularly when controlled by a single valve.

Figure 20 shows methods of applying three port valves.

a) shows the valve as a mixer in the flow with two inlets and one outlet. This gives constant volume at variable temperature

b) shows the valve as a diverter in the return with one inlet and two outlets. This gives constant volume at variable temperature

c) shows the valve as a diverter in the flow with one inlet and two outlets. This gives constant temperature with variable volume
d) shows the valve as a mixer in the return with two inlets and one outlet. This gives constant temperature with variable volume

The effect of the valve on the system depends on whether it is used as a mixer or a diverter and also on its position in the circuit.

Care must be taken when selecting a three port valve for use as a diverter. Fig. 21 shows typical valves. The "lift and lay" types are generally recommended for use as in Fig. 20(d) so that there will not be any sudden reversal of pressure to cause the valve to bang on to its seating. The rotary valve, Fig. 21(c), is not affected in this way. Manufacturers' recommendations on the location and connection of valves should always be followed.

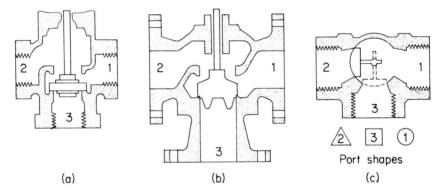

Fig. 21   Types of three-port valves
(a) lift and lay valve
(b) lift and lay valve
(c) rotary valve

§179   Roof Top Boiler Installations

The siting of boiler houses on the roofs of multi-storey buildings has become popular and offers architects and heating engineers a number of advantages over ground or basement floor locations.

New Buildings

The principal advantages include:

• no flues running the full height of the building, so giving increased floor space and simplifying internal planning, Fig. 22

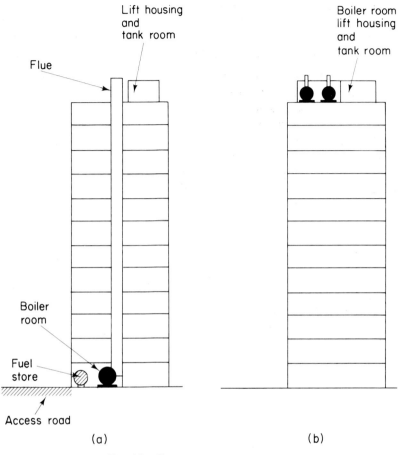

Fig. 22   Comparison of installations
(a) conventional
(b) roof-top

- no need for flue insulation to prevent condensation and the "overheating" of adjacent rooms is eliminated
- the roof top boiler room can be designed as a unit in conjunction with the water tank and lift housings
- the space not required in the basement or ground floor can be used for car parking or accommodation
- roof top boilers may easily be drained down since they stand above the heating circuits and require only a relatively small quantity of water to be removed when the boiler needs attention
- roof top boilers may be lighter and cheaper than basement or ground floor boilers which have to withstand high static pressures

• roof top boiler rooms are more easily ventilated and generally do not require mechanical extraction

Existing Buildings

The foregoing advantages apply equally to existing buildings. In addition, a roof top installation may offer the only solution to the problem of locating a boiler when one has not previously been fitted.

Modular boilers are particularly suitable since they may easily be transported by existing passenger lifts and manhandled into position.

Types of Boiler

Most conventional gas fired boilers are suitable for roof top installation. However, because the static pressure on the boiler is very low, light weight cast iron, sectional boilers which are unsuitable for basement locations, may be used. Shell boilers, which are designed for high pressures, can be used in either roof or basement locations.

Modular boilers are particularly suitable for roof top installation. They generally consist of a number of identical sectional boiler units, each with its own controls, coupled together to provide the desired output rating. Each individual unit is factory assembled and delivered to site. Several units are then connected together, sometimes by means of standard headers but more normally the headers are manufactured on site.

Control is usually by means of a step modulating system which maintains sufficient units in operation to meet the load. This eliminates cycling during periods when the full output of the installation

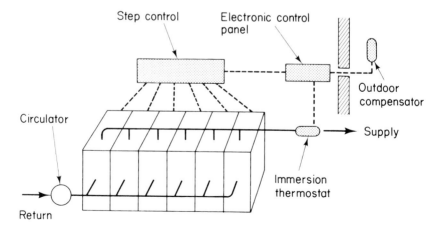

Fig. 23 Step modulating control system

is not required. A schematic diagram of the system is shown in Fig. 23.

When modular boilers are not used the roof top installation usually consists of more than one boiler. Multi-boiler installations may consist of boilers of similar or widely varying outputs used in any combination to provide the total heat output required.

Flues

Various arrangements are possible. Each boiler may have a separate flue, Fig. 24, or all may be discharged into a single flue either with or without induced draught or an air dilution system, section 182. The author understands that this practice is likely to be discouraged as unnecessary when BS CP332 Part 3 is revised.

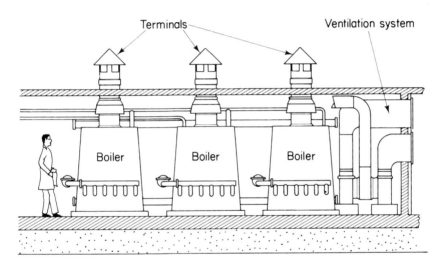

Fig. 24   Roof-top boilers with separate flues

General Considerations

Facilities must be provided for hoisting the boilers and associated plant into position. The hoists or cranes used in the construction of a new building may be retained to lift the central heating and water heating plant on to the roof. It must also be possible to replace the equipment subsequently without undue disturbance to the building or its occupants.

Because the boiler is at the top of the heating circuit, gravity circulation is not possible. All water flow through the system must be pumped.

The underside of the boiler bases must be thermally insulated to prevent the transmission of heat to the top floor of the building.

Anti-vibration measures are necessary to prevent noise and vibration from the boiler plant being transmitted to the structure.

To ensure a good standard of ventilation roof top boiler rooms should have at least two external walls.

§180   Water Heating

The sizing of boilers to meet commercial water heating demands has for many years been by rule of thumb. This has resulted in many boilers being oversized, sometimes by as much as 100%.

Because of the need to conserve energy, field tests are being carried out which should ultimately provide engineers with accurate methods of assessing the requirements of the various types of establishment. It is hoped by the Author that this information will eventually be published. Currently information is available in C.I.B.S. Guide Section B4.

Small quantities of water may be provided by multipoint storage water heaters similar to those used domestically but with higher outputs.

For large installations various types of boiler are used, often heating the water by a calorifier and also supplying central heating. A shell boiler with integral calorifier is shown in Fig. 25. The boiler and calorifier are at (a) and (b) and the pipework layout at (c).

With most large systems the draw offs are at a considerable distance from the source of hot water and a pumped secondary return is necessary. In multi-storey hotels a similar return would be necessary on each floor.

Where heating and hot water services are combined the boilers are generally controlled to give a variable volume of hot water at constant temperature. Whilst this is satisfactory for conventional boilers it gives problems in low water content modular boilers. At periods of little or no heating demand the mixing valve closes and all the heating water is recirculated. If, at the same time, the hot water demand is also minimal the calorifier will be bypassed and the boilers will be required to heat a very small volume of water. This is difficult to control and it is recommended that the heating and hot water services should be completely separate.

Independent systems, each with its own modular boilers, can be more effectively controlled and result in fuel savings.

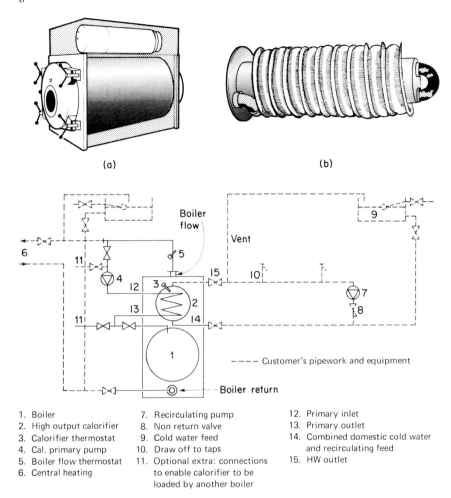

1. Boiler
2. High output calorifier
3. Calorifier thermostat
4. Cal. primary pump
5. Boiler flow thermostat
6. Central heating
7. Recirculating pump
8. Non return valve
9. Cold water feed
10. Draw off to taps
11. Optional extra: connections to enable calorifier to be loaded by another boiler
12. Primary inlet
13. Primary outlet
14. Combined domestic cold water and recirculating feed
15. HW outlet

(c)

Fig. 25   Shell boiler with calorifier
(a) boiler and calorifier complete
(b) calorifier
(c) pipework layout

§181   Changeover to Gas Firing

A large part of the expansion in commercial gas sales comes from the changeover of existing boilers to gas firing from other fuels.
  The subject of boiler changeover is dealt with in detail in the

British Gas publication, "Technical notes on changeover to gas of central heating and hot water boilers for non-domestic applications".

The notes apply to boilers with outputs above 44 kW (150 000 Btu/h) on hot water or steam systems providing central heating or hot water for non-domestic premises. The changeover of boilers in domestic premises is not at present recommended.

Early changeovers were predominantly from solid fuel. Up to 1973 this was about 93% of the total. Since then changeover has been principally from oil firing, up to about 80% of the total.

Advantage of Gas Firing

Although the decision to change to gas is basically a financial problem, the following factors may have some influence:

- there are no delivery or storage problems
- valuable space used for storage can be made available for other purposes
- no stock control necessary
- no advance payment for fuel
- no provision necessary for delivery vehicle access
- no ash or sludge removal
- lack of atmospheric pollution may allow shorter chimneys to be used
- reliability of fuel supplies
- gas boilers do not require an attendant except on very large installations
- easier to keep boilerhouse clean
- high thermal efficiency and controllability results in energy saving

Boiler or Burner Replacement

The simplest method of changing to gas firing is to fit a gas burner with its ancillary controls and to carry out appropriate modifications to the existing boiler. This is usually less expensive than replacing the boiler itself.

However, it depends on the age and condition of the boiler and also on its size and type. A boiler might better be changed if it is:

- too small or too big for the system
- in poor mechanical condition
- nearing the end of its life (about 10 years)
- of a type for which a suitable burner does not exist

Where there is any doubt it is usual for quotations for both courses of action to be requested by the customer. The difference in cost is least for the smaller boiler. Statistics show that boilers are more often replaced when more than 10 years old and less than 147 kW (500 000 Btu/h) output. Overall, boilers were replaced in one third of the cases.

Changeover

The first step is to examine the boiler and the installation. The points to note are as follows:

> Boiler —make, type, output rating
> —condition, age, faults
> —compliance with relevant standards and codes
> —compliance with AOTC requirements for insurance
> Installation —compliance with codes of practice
> —ventilation adequate
> —flue satisfactory
> —condition and faults
> —boiler size matched to heat demand
> —water treatment

The second step is to select a suitable gas burner. In the past many solid fuel cast iron sectional boilers were fitted with natural draught packaged burners. However, difficulties have arisen in setting up and operating some natural draught burners and it is recommended that they should only be used where quietness is an overriding consideration. Forced draught burners are considered by the author to be preferred.

The forced draught burner must have the correct output, fan pressure and flame shape to suit the boiler. Burner manufacturers should be consulted and most can recommend a suitable model.

Where an oil boiler has an equivalent gas version the same burner can usually be used.

Modifications to boilers

Solid fuel boilers require a number of modifications when changed over to gas firing. Loose firebars in good condition may be covered with refractory tiles. Alternatively, they may be replaced by pre-formed linings or a steel plate to support the tiles. If the boiler has water cooled firebars, the base should be filled up with a refractory material to just below the bars. The floor temperature beneath the boiler should not exceed 65°C (150°F).

The efficiency can usually be improved by restricting some of the passages between the combustion chamber and the flue. Some passages may be blocked to make the combustion products take a longer

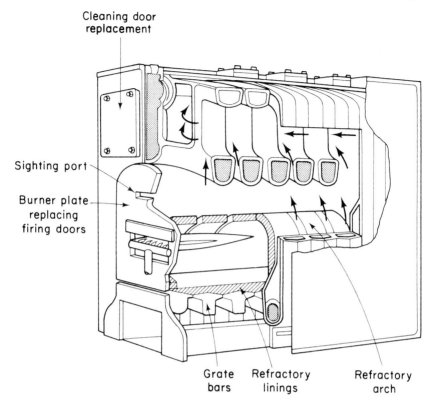

Cleaning door
replacement

Sighting port

Burner plate
replacing
firing doors

Grate   Refractory   Refractory
bars   linings   arch

Fig. 26   Sectional boiler modified for change-over to gas firing

route, always ensuring that combustion remains satisfactory. The use
of such methods will however also increase the fan horse power
requirements.

The inspection and cleaning doors must be replaced to make the
boiler pressure-tight. They may be covered with steel plates or be
replaced by composite panels which will fracture harmlessly if the
pressure rises excessively.

The sighting flap should be replaced by a sighting port and the
firing doors replaced by a steel plate on which the burners are
mounted. The extent of the modifications will vary with the particular boiler. Figure 26 shows a sectional boiler, modified for an
atmospheric burner. The refractory arch is used to cushion the boiler
against thermal shock.

Oil boilers generally require less modification. The cleaning doors
must always be replaced but if the refractory lining is in good condition and the gas burner fits on to the existing front plate, little else
may be required.

On changeover from oil the customer may wish to remove the oil tank so that the space may be put to other use. Removing an oil tank is a difficult job and it is best left to specialist firms who undertake this work.

Details of the procedure to be followed and the precautions necessary are detailed in the Department of Employment Technical Note 18, "The safe cleaning, repair and demolition of large tanks for storing flammable liquids".

Flueing and Ventilation

Ventilation requirements for gas fired boiler houses are given in BS C.P. 332, Part 3 (currently being revised for boilers 60 kW to 2 MW input) for installations up to 586 kW (2 million Btu/h) and in BGC notes "Combustion and ventilation air—Guidance notes for boiler installations in excess of 586 kW (2 million Btu/h)".

The British Gas notes have a waiver for boilers changed over to gas firing. If they previously operated satisfactorily on oil, there is no need to increase the ventilation rates should they be lower than specified by the notes. This is because gas requires approximately the same amount of air for combustion as oil.

It is not usual to line chimneys in commercial properties when changing to gas, however the flue should be swept clean.

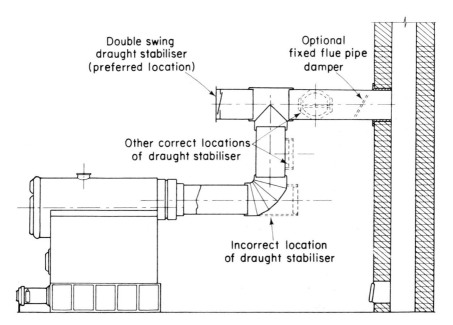

Fig. 27   Location of flue draught stabiliser

Changeover may give rise to dissimilar boilers firing into the same flue. Gas boilers should not share a flue with solid fuel boilers but may share with oil boilers. Natural and forced draught burners should not normally share a flue.

Because the resistance in a boiler fitted with a natural draught burner is about 0.5 to 1. mbar, a down draught diverter or sometimes a draught stabiliser is usually fitted to ensure adequate flue pull. A double swing stabiliser can relieve both excessive up-draught and down-draught and Fig. 27 shows possible locations for its installation. Flapping draught stabilizers can however lead to pulsating and noisy combustion.

FLUES FOR LARGE BOILERS

§182   Flues and Ventilation for Large Boilers

The design of flues generally is described in Volume 2, Chapter 5. In the case of the larger boilers, information on flue design may be found in the British Gas publication "Flues for Commercial and Industrial Gas Fired Boilers and Air Heaters".

The calculation of flue size is a complex procedure and only brief details are included here. The notes contain tables and nomograms which enable a satisfactory flue size to be determined. This is usually smaller than was obtained by previous methods but it has proved completely adequate.

Flue Systems

The principal systems to be considered are:

- natural draught flues
- fanned draught flues
- fan-diluted flues
- modular boiler flues

Natural Draught Flues

This system is usually fitted with a draught diverter which admits about 100% dilution and gives a CO_2 concentration in the secondary flue of about 4%.

Induced Draught Flue Systems

Many boilers use a combustion fan to provide forced or induced air to the burner. The fan is matched to the resistance of the appliance and the pressure at the flue spigot is usually zero. In some cases it is necessary for some flue pull to be provided. The flue is directly connected to the appliance so there is no dilution and the CO_2 concentration is about 8%. The flue could be natural draught or may be fitted with a fan.

With an induced draught flue system a fan is fitted into the flue to remove the combustion products. This system is normally used when:

- a natural draught system would not be satisfactory
  - —the flue has insufficient height to give the required pull
  - —the only possible flue run creates an exceptionally high resistance
- the natural draught system is unable to provide a satisfactory exit velocity for the flue products.

The major considerations, apart from size, when selecting and locating a fan are as follows:

- the fan may be required to withstand high temperatures, combustion products may be at more than 150°C (300°F)
- with any fanned system a safe start draught proving switch or air flow switch must be fitted to shut off gas to the burner in the event of the fan failing or the flue becoming blocked
- if the fan is fitted on the outlet of the boiler, the entire flue is under pressure and combustion products may leak into the premises
- if the fan is located just below the flue terminal, the entire flue is under suction and any leaks on the joints will only allow air to be drawn in.

Flue Design

The British Gas publication give two slightly different methods of sizing flues for natural draught or fanned draught systems. In brief, these methods are as follows:

Natural draught procedure:

- decide the route of the flue
- determine the vertical height to the terminal in metres
- determine the total length of the flue in metres
- determine the heat input rate of the appliance in kW
- decide on the CO_2 content of the flue gases

The minimum flue diameter may then be read off from tables for 4% or 8% CO_2 respectively. An extract is shown as Table 2 as an example. Corrections must then be applied for flue losses if less than 20% of the gross heat output of the appliance is lost in the flue and for flue

TABLE 2 Flue Diameter (mm) for 4% CO_2 Concentration

Appliance Heat Input, Q (kW)	Chimney Height, H_E (m)													
	2	3	4	5	6	7	8	9	10	12	14	16	18	20
50	162	149	144	139	136	134	132	131	130	129	128	127	127	126
75	196	180	172	166	162	159	157	156	155	153	152	151	150	149
100	225	205	197	189	184	181	179	177	175	173	171	170	169	168
150	273	248	237	227	221	217	213	211	209	205	203	201	200	199
200	314	284	270	259	252	247	243	239	237	232	229	227	225	224
250	350	315	300	288	280	273	268	264	261	256	252	250	247	246
300	383	344	327	313	304	297	291	287	283	277	273	270	267	265
350	412	370	351	337	327	319	313	308	303	297	292	289	286	283
400	439	395	374	359	347	339	332	327	322	315	310	306	303	300
450	465	418	395	379	367	358	351	345	340	332	326	322	318	316
500	490	439	416	398	386	376	368	362	356	348	342	337	333	330
550	513	460	435	417	403	393	385	375	372	363	357	352	348	344
600	535	480	454	434	420	409	400	393	387	378	371	365	361	357
650	557	499	471	451	436	425	416	408	402	392	384	378	374	370
700	578	517	488	467	452	440	430	422	416	405	397	391	386	382
750	598	534	505	483	467	454	444	436	429	418	410	403	398	394
800	616	551	521	498	481	468	458	449	442	430	422	415	410	405
900	653	583	550	527	509	495	483	474	466	454	445	437	432	427

length if this is significantly greater than the vertical height. The corrections may be obtained from other tables and nomograms in the notes.

Induced draught procedure:

- decide the route of the flue
- determine the total length of the flue in metres
- determine the heat input rate of the appliance in kW
- decide on the CO_2 content of the flue gases.

The volume of combustion products is then calculated or read from a nomogram and a flue diameter is selected to give a flue gas velocity of between 5 and 15 m/s (16 and 50 ft/s).

The pressure to be provided by the fan can be calculated from the resistances of the flue, the fittings and the appliance. A fan which will deliver the required volume against the static pressure drop is then selected from manufacturer's data sheets. If no such fan is available a new flue diameter must be chosen and the calculations repeated.

Fan Diluted Flues

These were described in detail in Volume 2, Chapter 5. They are generally used for small or medium sized commercial installations where it is not practicable to use a natural draught flue. A typical example is in ground floor shop premises with offices or flats above. Large boiler installations normally have natural draught flues.

The main considerations for fan diluted flues are as follows:

- the induced air should reduce the CO_2 content to below 1%
- the duct dimensions should give a flue gas velocity of between 6 and 8 m/s (20 and 26 ft/s)
- the inlet and outlet of the duct should preferably be on the same wall
- the outlet of the duct should be remote from the ventilation inlet
- the outlet should normally be at least 3 m above ground level although this is not always possible
- the fan should be fitted to the duct by flexible connections and have rubber mountings to minimise noise.

Modular Boiler Flues

The types of flue used may be:

- individual flues
- common flues with natural draught
- common flues with induced draught

Individual flues should be used, if possible. However, because of their improved appearance and economy, common flues are more frequently fitted.

With natural draught common flues not more than six appliances should be connected to the same horizontal header and not more than eight to the same vertical flue. If more than eight boilers are installed the flue should be fan assisted.

The vertical flue or chimney should preferably be straight and taken from the centre of the header or group of appliances. The minimum chimney height should be 2 m. The header should be the same size as the chimney for all its length and be horizontal. Headers should be fitted as high as the boiler house headroom will allow.

The connector flues link each appliance to the header. They should be at least 0.5 m high from the base of the draught diverter. They are normally connected into the header via a bend and in some cases may be fitted with flow restrictors. Fig. 28 shows a typical flue layout.

Natural draught flues are used where small numbers of boilers are installed. This is because, when some boilers are shut down, large quantities of cold air are drawn into the flue through their draught diverters. This limits the number of boilers which can operate satisfactorily since it reduces the pull of the flue.

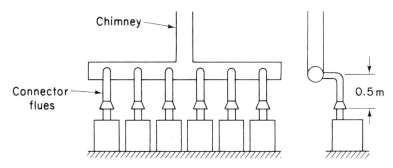

Fig.28   Modular boiler flues

With induced draught systems this problem does not occur. The volume of draught is dependent on the fan and is barely affected by the number of appliances in use. Fan-diluted systems may also be used where the installation is on the ground floor.

The sizing of natural draught modular systems is given in the British Gas publication. Briefly the procedure is as follows:

- decide the route of the common flue
- determine the vertical height from the top of the highest draught diverter to the top of the chimney

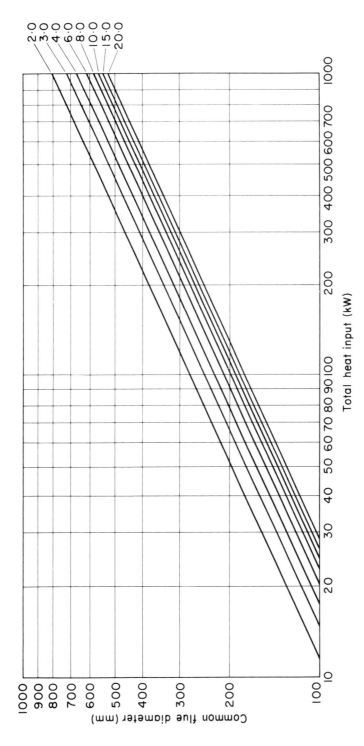

Fig. 29 Table giving common flue diameters

- determine the number of boilers and the total heat input rate in kW

The basic diameter may now be read off the graph Fig. 29. This is based on 3 appliances to each limb of the header, that is 3 appliances with an end chimney and 6 with a central chimney. A correction must be applied if 4 or more appliances are to be installed. The graph also assumes that the connecting flue has a 90° bend. If it enters the header vertically from below a further correction is required. These corrections may be obtained from the publication.

Materials for Large Flues

The main considerations when selecting the type of flue material are as follows:

- the inner wall of the flue must be capable of withstanding the highest temperature attained by the combustion products
- the flue wall must be resistant to condensation which is likely to occur when the appliance is started up from cold
- the flue should possess sufficient insulating properties or "thermal resistance" to prevent the continuous formation of condensation during normal running

The main consideration is the flue temperature and Table 3 gives the limiting temperatures of various materials.

Brick or concrete were the traditional materials used for tall flues.

TABLE 3 Temperature Resistance of Flue Materials

Material	*Limiting temperature*
Non-metals	
Brick	
lined with —acid resistant brick	up to 200°C (390°F)
or clayware	
—insulating brick	above 200°C (390°F)
Glass fibre reinforced plastic	about 250°C (480°F)
Asbestos cement—heavy duty	up to 260°C (500°F)
Concrete—in situ or pre-cast	about 300°C (570°F)
Metals	
Galvanised steel	up to 250°C (480°F)
Aluminium—high purity	up to 300°C (570°F)
Cast iron	up to 500°C (930°F)
Mild steel	up to 500°C (930°F)
Stainless steel (17% Cr)	up to 500°C (930°F)
Aluminised steel	up to 600°C (1110°F)

They were lined with insulating brick or clayware, flexible metal tubes are now commonly used.

The liner must be adequately sealed if the flue is under fan pressure. The space around the liner may be filled with loose insulating material, for example, vermiculite.

Asbestos cement flues were popular. They are obtainable in twin-wall form for additional insulation. All sockets should face upwards and be satisfactorily sealed. Each section of pipe should be supported.

Metal flues are now more commonly used. Most metals are satisfactory if there is no condensation. Where condensation does occur, aluminised or stainless steels are the most resistant. Metal flues are also available in twin-wall form and, in some cases, insulating material is introduced into the space between the two walls.

Some metal flues, produced by specialist manufacturers, consist of a welded tube inside a self-supporting or guyed windshield. Several items of plant may be flued together within a common windshield, the tubes, being insulated with loose fill material.

With all flues, facilities must be provided for their inspection and renewal when necessary.

Heat Loss from Flues

To maintain the maximum flue pull and to avoid condensation the heat losses from the flue gases must be kept to a minimum. So flues are run inside the premises, where possible, or insulated when run externally. The object is to maintain the inner wall temperature above the dewpoint temperature of the combustion products. This is about 60°C (140°F).

Where this is not possible and condensation is inevitable the flue should be designed so that the water can flow freely to a point from which it can be drained to a gulley.

§183   Ventilation Requirements

Notes giving guidance on the combustion and ventilation air requirements for boilers with heat outputs above 586 kW (2 million Btu/h) have been published by British Gas.

The notes are based on the following assumptions:

- the maximum excess air with a natural draught burner is 0.04 m³/s (5000 ft³/h) per 293 kW (1 million Btu/h) of boiler output and 0.024 m³/s (3000 ft³/h) per 293 kW (1 million Btu/h) of output with a forced draught burner
- the heat released into the boiler house is 3% of boiler output

- heat loss through the boiler house is $51.5 \, W/m^3$ (5 Btu/h per ft^3) of its volume
- the minimum boiler house volume is $21 \, m^3$ per $293 \, kW$ (750 ft^3 per 1 million Btu/h) of boiler output
- the size of the ventilators should allow an approximate temperature rise in the boiler house of 28 deg. C (50 deg. F)

If the assumptions are not met, the ventilation provided should be adequate to maintain the temperature within the boiler house below $32°C$ ($90°F$).

Boilers with Natural Air Supply

The main considerations are:

- ventilation grilles should open directly to outside air
- they should be located so that they cannot be easily blocked or flooded
- high level grilles should be as near as possible to the ceiling
- where only high level openings are possible it may be necessary to carry combustion air to floor level by means of a duct
- grilles should be designed to avoid setting up high velocity air streams in the boiler house
- where a boiler house is exposed, grilles should be sited on at least two sides and preferably on all four

The sizes of ventilation grilles are given in Table 4.

TABLE 4   Minimum Ventilator Free Area for Natural Air Supply*

Heat Output kW (Btu/h)	Location	Area m^2 (ft^2)
586 to 1025 (2 million to $3\frac{1}{2}$ million)	Low level	0.65 (7)
	High level	0.325 (3.5)
above 1025 ($3\frac{1}{2}$ million)	Low level	0.634 m^2 per 1000 kW of boiler output (2 ft^2 per 1 million Btu/h)
	High level	0.317 m^2 per 1000 kW of boiler output (1 ft^2 per 1 million Btu/h)

*B.S. CP 332 Part 3 is currently being revised.

Boilers with Fanned Air Supply

The following points should be noted:

- two fans may be used, one to supply air the other to extract, the extraction rate should not exceed the supply rate

TABLE 5 Air Requirement for Fanned Air Supply*

Air Requirement	Air Flow Rate in³/s per 1000 kW of Boiler Output (ft³/min per 1 million Btu/h)			
	Natural draught burner		Fanned draught burner	
	Draught diverter	Flue stabiliser	Direct flue connection	Flue stabiliser
Minimum combustion and dilution air	0.89 (550)	0.6 (370)	0.41 (250)	0.54 (335)
Minimum combustion dilution and ventilation air	1.46 (900)	1.31 (810)	1.19 (735)	1.26 (775)
Extract air	0.57 (350)	0.71 (440)	0.78 (485)	0.71 (440)

*B.S. CP 332 Part 2 is currently being revised.

- if only one fan is used it should supply air and extraction should be by natural ventilation
- supply air should enter at low level
- natural air vents should be at high level and sized as in Table 0.
- an air flow switch must be fitted to shut off the gas to the burner should the fan fail

The volume of air required for fanned air supply is given in Table 4.

Boilers Changed Over from Oil to Gas

Where boilers have been changed over from oil to gas firing and where the air vents have proved adequate for the oil fired boiler, there is generally no need for them to be altered, even though they may be smaller than those recommended by Table 4 and 5.

Index

Printed in Great Britain by The Anchor Press Ltd
and bound by Wm Brendon & Son Ltd, both of Tiptree, Essex